GALILEO
and the
Scientific Revolution

Laura Fermi
and
Gilberto Bernardini

Dover Publications, Inc.
Mineola, New York

ACKNOWLEDGMENTS

Several persons have assisted us in the preparation of this book, and to them we wish to express our gratitude. To Professor Cyril Stanley Smith of the University of Chicago for his help in editing our translation of the "Little Balance," for his note on it, and for his unstinted advice. To Professor I. Bernard Cohen and Dr. Edward Grant of Harvard University, and to the Reverend Ernan McMullin, C. S. C., for their criticism of the entire manuscript. To several members of the Physical Science Study Committee for their assistance in preparing the manuscript and their many valuable suggestions.

This book contains excerpts of Galileo's works and correspondence and of other documental material. We have quoted some of the excerpts from existing translations by Stillman Drake, by Giorgio de Santillana, and by Henry Crew and Alfonso de Salvio. We have also translated directly from the Italian National Edition of Galileo's Works.

Bibliographical Note

This Dover edition, first published in 2003, is an unabridged republication of the 1965 Fawcett paperback edition of the work first published by Basic Books, Inc., New York, in 1961.

Library of Congress Cataloging-in-Publication Data

Fermi, Laura.
    Galileo and the scientific revolution / Laura Fermi and Gilberto Bernardini.
       p. cm.
    Originally published: New York : Basic Books, 1961.
    Includes bibliographical references and index.
    ISBN 0-486-43226-2 (pbk.)
    1. Galilei, Galileo, 1564–1642. 2. Astronomers—Italy—Biography. I. Bernardini, Gilberto. II. Title.

QB36.G2F43 2003
520'.092—dc21
[B]
                                                 2003050231

Manufactured in the United States of America
Dover Publications, Inc., 31 East 2nd Street, Mineola, N.Y. 11501

CONTENTS

INTRODUCTION

Galileo ... Who Was He?

ON A SEPTEMBER DAY in 1581 a 17-year-old Italian boy registered at the university in the Tuscan city of Pisa, in central Italy. He had come from Florence, some 50 miles east, to study medicine, the only profession in which a boy could hope to make a decent living at that time. According to the current custom, his name was recorded in the university register in Latin: "Galileus Vinc^u Galilei Flor^s Art^a"—Galileo, son of Vincenzio Galilei, Florentine, Student of Arts.

We cannot know for certain what the boy Galileo was thinking when he registered, but we can make some guesses. He was probably happy to be back in Pisa, where he was born and where he had spent the first ten years of his life. Pisa was a beautiful city, encircled by medieval walls and towers and bisected by the Arno River, which brought the trading vessels from the Tyrrhenian Sea, a few miles away. The city was proud of its white marbled monuments, the most famous of which was then, as it is now, "the Leaning Tower."

Undoubtedly the boy felt prepared to enter the university, knowing Latin and Greek—languages which European scholars studied then, as they still do. Besides, he had studied logic with a monk in a quiet monastery up in the mountains near Florence, among woods of pines and birches; he could draw and paint; and from his father, an accomplished musician, he had learned to play the lute. This was a well-rounded education in a time when children did not always go to formal schools but often studied, as Galileo had done, with their parents and with private tutors, both lay and religious.

7

He was probably full of expectations, because exuberance, optimism, and lively imagination were among his traits. If he hoped to become famous in his lifetime and to have statues built to him after his death, then his hopes were fulfilled. Memorials of all kinds now honor him in Italy and wherever science is appreciated, and in a hall of the very building where he registered on that September day there stands today a statue of him.

If he thought that he would win fame as a physician and save many lives or rid humanity of disease, then he was wrong. He never became a physician, because mathematics diverted him from his original purpose. The power of mathematical reasoning impressed him so much that he became interested in applying it to the events that occur in nature, to the way in which things move and fall. This study of motion he began in his own way, with a keen sense of observation and objective judgment. What he learned turned him gradually against the well-established tradition. His teachers studied science only in books; scholars who studied it from nature stopped too often at the *observation* of natural phenomena and did not make scientific experiments, in the sense accepted in modern science. Galileo instead *invented* experiments, considered them a necessary regular practice, and came to regard them as the criterion in discriminating between the unlimited possibilities of human imagination and the reality of the physical world. To exclude subjective influence and attain complete objectivity, he described natural phenomena and the results of his experiments in the rigorous language of mathematics.

He taught his way of performing and interpreting experiments to his pupils and to the nonscientists. In this also he was an innovator: He went against the longstanding habit of writing science in Latin for the benefit of the few learned men in foreign countries and wrote instead in Italian, in a pleasant, popular style which almost any literate person could understand. His books, received with great interest and translated into other languages, were and still are models for those who recognize the power of science and its essential place in culture.

Nature had been generous to Galileo. It had given him not only a good mind but also skillful hands. What his acute mind could design, his hands could build. Thus he built a telescope which extended the limited powers of human senses in ways that no one had foreseen. He saw a universe "enlarged a hundred and thousand times from what the wise men of all past ages had thought"; he saw new and extraordinary phenomena in the sky. The thorough study of these phenomena led him gradually to conclusions which went against the accepted principles of his time. To defend his conclusions and with them the freedom of thought in science, he engaged in a long battle which brought persecution down on him and forced him to spend the last years of his life in seclusion. But, as Max Born, a scientist of our time, has said: "The scientific attitude and methods of experimental and theoretical research have been the same through the centuries since Galileo and will remain so." *

* Max Born, (Nobel Prize, 1954), *Physics in My Generation,* Pergamon Press, 1956.

1. Student in Pisa

GALILEO GALILEI was born in Pisa on February 15, 1564. But a date has no meaning unless it is related to well-known events, and to build a frame around that distant February day we shall recall some historical facts.

When Galileo was born, only 72 years had passed since Columbus's discovery of America. Not until the next year would Europeans succeed in establishing a settlement in the land that is now the United States, at St. Augustine, Florida. The first successful colony in Virginia was to start in 1607, when Galileo was 43 years old, and the *Mayflower* was to anchor at Plymouth when he was 56. He was a baby of two months when Shakespeare was born in England, under the reign of Queen Elizabeth.

Italy was divided into many independent states. Both Florence and Pisa were in the Grand Duchy of Tuscany, in the central and western part of Italy, ruled by a Grand Duke of the Medici family. Tuscany, and Florence in particular, had been the center of the Italian Renaissance, the intellectual movement of the fourteenth to the sixteenth century which had ended the rigid, conservative culture of the Middle Ages and prepared the way to the enlightened thought of our modern Western civilization. Evidence of the extraordinary degree to which all the arts had flourished in Florence were its famous palaces, churches, and paintings that many tourists still go to see.

The artistic period of the Renaissance was coming to an end. Only three days after Galileo was born, Michelangelo, the greatest of all Florentine artists, died in Rome. Western civilization, after a long period of joyful expression, of colors and forms, was getting ready for inner

11

thinking, for criticism and science. Although art had reached maturity, science was still in its infancy. The official science, taught by famous teachers in all European universities, studied by thousands of students, still lacked the open-mindedness that had become well established in the arts.

The Galileis were a noble family of Florence who had lost most of their money and had some difficulty making ends meet. Vincenzio, Galileo's father, was an intelligent, well-educated man and an able musician. Not only did he play and teach music, but he also wrote several highly appreciated treatises on the theory of music, some of which are still preserved. In order to make a living, Vincenzio became a cloth merchant and moved temporarily to Pisa, which was then a better trading center than Florence. He and his family returned to Florence when his oldest son, Galileo, was ten years old. He had several other children, some of whom died young, and three of whom we know something about: two daughters, Virginia and Livia, and a son, Michelangelo.

Vincenzio directed Galileo's early education. The boy learned easily and eagerly and must have shown unusual intelligence and alertness, because despite the family's difficult financial condition the father decided that Galileo should go on with his studies at the university. This meant that the boy would live away from home, since the university of the Grand Duchy of Tuscany was in Pisa. If the decision on what to study had been left to Galileo, he would have taken up drawing and painting, he told his friends many years later. But his father had in mind a more remunerative position for him, and so he entered the University of Pisa as a student of medicine.

ARISTOTLE'S AUTHORITY

One of the fundamental courses he was required to take was the philosophy of the great Greek Aristotle, who lived in the fourth century before Christ (384-322 B.C.). In Galileo's time the word "philosophy," which in Greek means "love of wisdom," had a much broader meaning than it has now, and it included the study of all natural phenomena. Aristotle wrote a large number of books in which he presented the entire philosophical and scientific knowledge of his time with a broadness of scope, power of synthesis, and originality that have never been matched. He also founded a school which had many followers and exerted a great influence on the Greek and Latin cultures.

Of the huge bulk of Aristotle's work, Galileo studied mainly the writings on logic, on motion, and on the structure of the universe. In following Galileo's steps, we shall have occasion to show some of Aristotle's views. Here we shall say only a few words about Aristotle's conception of the world.

Many philosophers before Aristotle had speculated about the appearance of the skies—the myriads of fixed stars, the apparently complicated paths of the moving stars (which we now call planets), the relation of our earth to sun and moon. Many had tried to fit what they had seen into coherent systems, often using complicated geometrical and mathematical schemes to explain the motion of celestial bodies.

Aristotle did not believe that mathematics was necessary to explain change and motion. His natural philosophy was quite primitive and strongly influenced by the immediate appearance of phenomena. He saw, for instance, change and deterioration on earth but *never* in the sun and stars. He did not consider the possibility that change and deterioration might take place in the sun and stars as on the earth but might not be seen because the sun and stars are so distant from our earth. He stated that the sun and stars were perfect and unchangeable.

In Aristotle's universe the earth was at the center and all celestial bodies turned around it, moving on transparent spheres. They moved with uniform circular motion, the most perfect form of motion, according to the ancients. The universe was divided in two regions: a sublunar region, below the sphere on which the moon moved, and an outer region, from the sphere of the moon

to that of the fixed stars. In the sublunar region everything was made of four elements: air, water, earth and
fire; everything was corruptible and changeable and there
could be death and birth. In the outer region, containing
the planets and stars, everything was made of ether and
was perfect and immutable.

Since Aristotle's time, the science of cosmology and
astronomy had of course progressed, but most philosophers still accepted only those results of observation that
they could reconcile with Aristotle's system of the universe.

Young Galileo began to study Aristotle's texts diligent-
ly, as we learn from some Latin commentaries, *Juve-
nilia,* which he wrote while a student at Pisa. As he read,
he came gradually to realize that in Pisa he really did not
have several teachers but only one teacher, this philosopher Aristotle, who had been dead for 18 centuries. On
all questions concerning the study of nature, Aristotle
was the source of authority—either in his own writings
or in the commentaries of those who had gradually modi-
fied or altered the Aristotelian point of view.

Although at first he went along with the trend, Galileo
spent most of his adult life in a struggle against the Aristotelian tradition and the men who accepted it blindly. A
few words about the meaning of Aristotle's teachings
and his authority will help to measure Galileo's stature
and to explain the difficulties he faced in breaking from
tradition.

Aristotle had a great following in the Greek and Latin
worlds, but after the fall of the Roman Empire most of
his works were lost to Western civilization and were for-
gotten for several centuries. Only in the Byzantine world,
the cultural heir of ancient Greece, did Aristotle's works
survive. They were soon translated from Greek into
Syriac, and, after the Arabs conquered Syria, from Syriac
into Arabic. In the twelfth and thirteenth centuries these
Arab translations, and Arab interpretations of Aristotle's
writings, were translated into Latin. In all these wander-
ings Aristotle's thought had suffered changes and distor-
tions. He would hardly have recognized some of the views

attributed to him. Yet it was these dubious versions of his writings that revived Western interest in him.

Shortly afterwards several of Aristotle's Greek texts were found in Constantinople (the ancient Byzantium). Under the impulse of St. Thomas Aquinas, one of the Doctors of the Church, Catholic monks translated them directly from the Greek into Latin. St. Thomas himself studied Aristotle's texts and wrote commentaries which over the centuries profoundly affected the Church's thinking. There were important ideas in Aristotle's works which St. Thomas could not explicitly identify with the religious dogma of the Church, but the harmonious and predetermined construction that Aristotle attributed to the universe found a perfect place among these ideals. St. Thomas gave Christian form to Aristotelian thought. Greek perfection, an abstract, geometrical, and aesthetic idea, was transformed into the perfection of God's works. Natural phenomena, and especially the eternal motion of celestial bodies, became a reflection of such perfection, and their ultimate purpose became that of glorifying God, His will and His inscrutable ends.

Over the years Galileo criticized and refuted many of Aristotle's views. In time he became one of the most relentless demolishers of Aristotle's doctrines, and perhaps the most effective. However, he was neither the first nor the only one in his time, for the sixteenth and seventeenth centuries were characterized, in Europe, by a process of profound revision of natural philosophy. In little less than a century this process changed the image that man had of the universe more fundamentally and rapidly than ever before in history. But even before all this started, some serious criticism had been raised against the solidly established Aristotelian structure. At the University of Paris in the fourteenth century, for instance, the Frenchman Jean Buridan and others had not accepted Aristotle's explanation of motion. We know that the motion of a body originally at rest is caused by forces. To the Greek philosopher a force (or more precisely the cause of velocity) could not be separated from a being, man or God, who somehow perceived its intensity. Aristotle also believed that things moved only when a force acted on them; that if the action of a force was not evident on a body in motion it was because the particles of surrounding air propagated and maintained this action: When a body received a push it started to move, leaving a vacuum behind; into this vacuum rushed particles of air pressing the body on, because nature, as most ancients believed, "abhors a vacuum." Buridan tried to find a more rational explanation of motion, and in this he was one of the most eminent among Galileo's precursors.

Any uncritical acceptance of one man's opinions was not consistent with Galileo's independent spirit and the broad-minded education he had received at home. His father, Vincenzio, was a man of liberal views and was used to expressing them openly. At about the time Galileo entered the university, Vincenzio published a *Dialogue of Ancient and Modern Music* in which he made one of his characters say: "It appears to me that they who in proof of any assertion rely simply on the weight of authority, without adducing any argument in support of it, act very absurdly. I, on the contrary, wish to be allowed freely to question and freely to answer you without any sort of adulation, as well becomes those who are in search of truth."

The perceptive son of a man who could write this way was bound gradually to realize that his teachers in Pisa acted exactly in the manner that his father considered absurd. There were in Pisa, as in other European universities, a great rigidity and a mental inertia that may now seem surprising. They were especially noticeable in anything concerning science, including medicine, which was then mixed and confused with philosophy, theology, and astrology. The bold freedom of thought of the Renaissance had remained within the vast but definite limits of magnificent works of art. It had not yet surmounted the barrier of the traditional philosophy, which had by then become strongly tied to religious belief.

Trends weakening these barriers started to appear in the sixteenth century as a natural consequence of the Renaissance intellectual innovations. Young Galileo, being the son of a man who was both a musician and a mathematician, must have been especially sensitive to these new trends and receptive to innovations. Since his very respectable teachers provided no innovations, he read directly in "this immense book that nature keeps open before those who have eyes in their forehead and brains."

And thus he made his first discovery, according to his most devoted pupil and earliest biographer, Vincenzio Viviani (1622-1703). But in relating episodes of his teacher's life, Viviani, who met Galileo only when Galileo was already a very old man, often embellished them, and so he may have done when he described the first discovery his old teacher had made when a young student. Yet this story is of great significance.

SWINGING LAMPS AND PENDULUMS

One day in the Cathedral of Pisa, says Viviani, Galileo was watching a lamp which had been moved from its rest position and was swinging from the ceiling. Probably guided by his musical education, he observed a rhythm in the swings of the lamp. It seemed to him that the lamp always took the same time to go from one end of its swing to the other, although its successive swings were gradually decreasing in amplitude. The reason for this, he thought, could be that the greater velocity of the lamp during the first swing made up for the longer path which it had to travel. How could he find out whether all swings actually took the same time?

He thought of measuring the times of swings. But there were no wrist watches then, and had a clock been available it would have been of little help, because clocks were not accurate enough to measure short times. Galileo, the student of medicine, put his finger on his pulse and counted the beats. Every oscillation of the lamp lasted the same number of beats.

Not satisfied with this proof, he went home—he lived

with relatives in Pisa—and tied two small balls to two strings of exactly the same length. He swung one ball a certain distance from the rest position and the other a different distance. While he watched one pendulum, a friend watched the other and they both counted the oscillations. They found that both pendulums made the same number of oscillations in the same time. Even balls of different weight oscillated with the same period, provided they were hung on strings of the same length. Thus Galileo discovered the law of *isochronism* (equality of time) of small oscillations.

Historians of science have ascertained that the famous lamp in the Cathedral of Pisa, which is still called Galileo's lamp, was hung three years after he is said to have watched it swing. But it could have been another lamp in the same church. Galileo himself did not leave in writing a description of his earliest experiments. He referred, however, to observations, "especially in churches where lamps, suspended by long cords, had been inadvertently set into motion." Besides, both in a letter of 1602 and in the greatest of his books, *Two New Sciences,* he described experiments which fit into a part of Viviani's story, without saying when he performed them. In *Two New Sciences* he says:

". . . I took two balls, one of lead and one of cork, the former more than a hundred times heavier than the latter, and suspended them by means of two equal fine threads, each four or five cubits long. Pulling each ball aside from the perpendicular, I let them go at the same instant, and they, falling along the circumferences of circles having these equal strings for semi-diameters, passed beyond the perpendicular and returned along the same path. These free goings and returnings repeated a hundred times showed clearly that the heavy body maintains so nearly the period of the light body that neither in a hundred swings nor even in a thousand will the former anticipate the latter by as much as a single moment, so perfectly do they keep step."

Clearly, in this crude experiment these two pendulums could not have kept perfect step in "a thousand swings."

In other experiments also Galileo claims more precise results than he could have obtained with his apparatus. But Galileo had a remarkable ability to judge the effects of unessential factors on the results of his experiments: resistance of air and friction of the string would stop a pendulum like Galileo's long before it had made "a thousand swings." Galileo knew that if the resistance of air and friction were removed, if all experimental conditions were ideal, the pendulum *would* perform "a thousand swings" all lasting the same time. (His "a thousand times" may mean "many times"—in Italian the word *mille* [thousand] is still used in the sense of "many.")

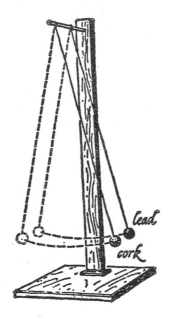

Galileo's Pendulum Experiment. The cork ball and the lead ball took the same time for one complete oscillation when they were suspended on strings of the same length.

In the light of these considerations Viviani's story acquires significance. It shows the method that Galileo followed in his experiments. He gathered information through the senses *observing* the phenomenon (the swinging lamp) and tried to understand its essential features. Then he went further. He formulated an *hypothesis:* perhaps all oscillations, of great and of small amplitude, lasted the same time. After a first check of this hypothesis (comparing the times of swings with his own pulse beats), he devised an experiment. The experiment *reproduced the essential elements of the phenomenon under controlled and simplified conditions.* The complicated shape of the lamp might have influenced its motion, so he replaced it with bodies of the simplest shape, round balls, hung on strings whose length could be measured exactly. Only after obtaining the same result over and over again under these controlled conditions would Galileo be satisfied that the hypothesis was valid.

This is an example of a rigorous scientific approach, of a classical experimental pattern. The experimental method was not new in science, but in what we now call physics it had been used only occasionally. Greek scientists had performed a few experiments, for instance on floating bodies and on the propagation of light. In the Middle Ages, also, scientists had occasionally made experiments, especially in optics and magnetism.

Galileo was among the first, if not the first, to use experiment in the study of motion. More important, he came to consider experiments as the only sure way of checking an hypothesis against facts and discovering truth. Sometimes he performed the experiments he devised; other times he carried them through in his mind. Through his faith in it the experimental approach was to become the solid foundation on which modern science grew.

According to Viviani's story, at an early age Galileo was already interested in what we now call physics. If we are to try to reconstruct the development of his complex and unusual personality as faithfully as possible, we may find it significant to know how he actually broke away from the study of medicine, on which his father had started him.

FROM MEDICINE TO MATHEMATICS

He had reached his third year at the university and the age of 19, with no training at all in mathematics, when he met Ostilio Ricci, the court mathematician of the Grand Duke. At the court in Florence the Grand Duke reared and educated 60 young pages from noble European families and gave them the best available teachers. When the Grand Duke spent some time in Pisa, as he often did, his court, including pages and teachers, followed him. It was at one of the court's temporary stays in Pisa that Galileo, on his father's insistent advice, went to hear Ostilio Ricci explain the geometry of Euclid. Geometry so fascinated Galileo that he abandoned the study of medicine.

Galileo's own son, writing after his father's death, related this decisive event in Galileo's life in these suggestive terms:

"Galileo, occupied in the study of medicine, for some time showed himself disinclined to mathematics, although his father, who was highly proficient in it, exhorted him to [study] it; finally to satisfy his father he applied his mind to it; but no sooner did he begin to taste the manner of demonstrating and the way of reaching the knowledge of truth, than, abandoning any other studies, he gave himself entirely to mathematics."

In 1586 he left the University of Pisa and went back to his family in Florence.

2. The Young Teacher

IN FLORENCE, Galileo applied himself to geometry and from the study of Euclid he soon passed to that of other ancient mathematicians, especially the Greek

Archimedes (287-212 B.C.), famous for his studies of the behavior of bodies immersed in water. Galileo soon recognized that if Aristotle was the greatest philosopher of antiquity, Archimedes was its greatest scientist and one of the greatest mathematicians of all time. He was the first to apply what we now know as calculus to the determination of areas and volumes.

It is certain that Galileo found in Archimedes a teacher from whom he learned the power and the wide intellectual range of mathematical reasoning, which imposes on imagination no limit but logic.

Galileo studied earnestly, as is proved by two works he wrote in this period. One, in Latin, which contains theorems to determine the center of gravity of solid bodies, made him favorably known among mathematicians of his time. He did not publish it then but circulated it in manuscript copies. Many years later he included it in the book *Two New Sciences*, which we mentioned earlier.

The other work is called *The Little Balance* and was written in Italian. Perhaps Galileo chose to write it in the language of the day so that Italian artisans, who did not read Latin, would be able to build and use this instrument.

ARCHIMEDES AND THE BALANCE

In *The Little Balance* Galileo told how he had read the story of Archimedes and Hiero, King of Syracuse in Sicily. The King wanted to know whether his crown was of pure gold or whether his jeweler had cheated him and mixed gold with silver. Archimedes, according to the story, weighed the crown, took an equal weight of gold, and immersed both pure gold and crown, one after the other, in a basin filled to the brim with water. The amount of water spilled out was greater when he immersed the crown than when he immersed the pure gold. This showed that the crown was of greater volume than the piece of gold. Since both were of the same weight, the crown was not of pure gold but contained silver, which is lighter than gold.

Galileo rightly believed that Archimedes could not have detected the small difference in the density of pure gold and of a mixture of gold and silver in the way told in the story. He thought that a more accurate method was needed, and to achieve this he built an instrument, the "little balance," which we now call hydrostatic balance. This instrument, an improvement on earlier ones, is based on Archimedes's principle that a body immersed in water receives a push upward equal to the weight of the water which it displaces. A mixture of gold and silver immersed in water displaces more water than an equal weight of pure gold. Therefore, it receives a greater push upward. The little balance measures differences in pushes and therefore differences in density.

In the Appendix we give a full translation of this work.

In his stay in Florence, Galileo gave some public lectures and private lessons both in Florence and in the nearby city of Siena. He taught not only mathematics but also literary subjects. Sometimes he put the two together, as in certain lectures on Dante's great poem *The Divine Comedy,* when he applied mathematics to calculate the size of hell and of its various parts.

There can be no doubt that these activities, which indicated that Galileo did whatever he could to make money, were poorly paid. He was certainly not self-sufficient, and very far from being able to help his family, as he might have done had he become a physician. We are not surprised that in this period he was busily looking for a stable position. According to the current practice, he repeatedly enlisted the influence of family friends and tried to obtain a chair of mathematics at the universities of Pisa and Bologna and at the court of Tuscany in Florence. At last, in July 1589, at the age of 25, he was appointed lecturer on mathematics at the University of Pisa. The position of lecturer in those times is comparable to that of one of our young professors at the beginning of his career.

ON MOTION

In the fall of that year Galileo moved to Pisa, and while he taught at the university he went on studying. His salary was small, smaller than that of most of the professors, who were older than he. Whenever he could not deliver a scheduled lecture because of illness or other unavoidable cause, the university cut his salary in proportion. Not even at this time could he be of much help to his family.

In this period he resumed the study of motion and wrote a short book in Latin, the *De Motu* (*On Motion*). This book seems to be a course of lessons elaborated into the first draft of a textbook. Galileo left it in manuscript form, and it was not published in its entirety until three centuries later.

In *De Motu,* Galileo tried to disprove some of Aristotle's main views about motion. One of these, for instance, was the assertion, which Buridan had rejected in the fourteenth century, that motion, in the absence of the direct action of a force, is maintained by the medium in which it takes place.

We cannot know whether Galileo was acquainted with Buridan's work. There were no scientific journals then, but only printed books, and papers were usually circulated in a limited number of manuscript copies. There was not yet the custom of always indicating the sources of information, and it is often hard to judge whether there was any relationship between two works on the same subject. But in any event, Galileo went much further than Buridan in the study of motion.

De Motu represents Galileo's first step in his systematic, independent study of natural phenomena. We shall come back to it in a later chapter, in which we shall try to explain the evolution of Galileo's ideas on motion. *De Motu* is of interest to us here as a point of reference in Galileo's intellectual development. It shows that when Galileo wrote it, in 1590, he had already started the deep process of original thinking and critical

revision of Aristotelian principles which lasted throughout his life.

FALLING BODIES

At about this time Galileo tried to disprove in a dramatic way one particular statement of Aristotle's, if we
can believe his biographer Viviani. Aristotle said that
when bodies of the same material but of different weights
fall freely, they fall with speeds proportional to their
weights. To show that this is not true and that lighter
and heavier objects fall with approximately the same
speed, Galileo, so the story goes, climbed to the top of
the Leaning Tower of Pisa while the entire body of students and teachers gathered in the square below to see
the demonstration. From the top of the tower Galileo
let go two bodies of the same substance but of different weights at exactly the same time. They reached
the ground together.

This story is probably not true. Galileo was not a
modest man and always sought recognition for his work.
It is unlikely that if he really had performed this experiment he would not have explicitly talked of it in one
of his numerous writings. Besides, it now seems certain
that the Dutch mathematician Simon Stevinus (1548-
1620) performed the experiment in Holland.

What is true and told in *De Motu* itself is that Galileo,
without actually performing an experiment, had reached
the conclusion that two bodies of the same substance
must (if air resistance is disregarded) fall the same distance in the same time.

A similar conclusion had been reached only a few
years before by G. B. Benedetti and described in a
book published in 1585, which Galileo probably read.
Galileo (and Benedetti) reasoned that if two homoge-
neous bodies were falling side by side it could make no
difference whether they were separate or united to one
another to form a larger body.

Often Galileo reached important conclusions by devising an experiment which he could not, or would not,

physically perform. Step by step, by logical deduction, he would carry the experiment through in his mind to its inevitable result. Mental experiments were in use long before Galileo's times, and scientists now call them "thought experiments." For their number and brilliance Galileo's "thought experiments" have become very famous.

Scattered throughout *De Motu,* reasonings and considerations of the type of those on falling bodies clearly mark Galileo's departure from traditional philosophy.

FROM PISA TO PADUA

Galileo taught only three years in Pisa. During his stay he kept trying to obtain the chair of mathematics at the University of Padua, which indicates that he was not satisfied with his position in Pisa. It is quite likely that he had begun to feel ill at ease among his colleagues, most of whom were staunch supporters of Aristotle's opinions, both right and wrong, as they had been when Galileo was a student. He probably irritated some of them with his self-assurance and ability to make fun of others. It is at about this time that he wrote a poem to ridicule a new regulation requiring all professors to wear their togas not only in classrooms but also when strolling in town or along the banks of the Arno River. However, the description of a young combative Galileo, anti-Aristotelian all the way through, is the product of Viviani's dramatic emphasis, of other biographers' interpretations, and of popular fantasy. Young Galileo was certainly witty and lively and enjoyed making fun of many contemporary university customs. With his colleagues "he was used to disputing fiercely about many a nice tasty thing." But from the terms in which some of his colleagues in later years recalled these "fierce" disputes, we are drawn to think that most of the time they were conducted in a courteous and highly entertaining manner.

It is likely that Galileo tried to get a better salary in order to meet the increasing needs of his family. One of his sisters was married in this period, and although he could not possibly have set a penny aside he pledged himself to give her a dowry. In those times a girl was never self-supporting because she never worked outside the home; she could not even make a decent marriage unless her family gave her a considerable sum for a dowry.

The conditions of Galileo's family worsened when, shortly after his sister's marriage, his father died and Galileo became the main support of his mother, younger brother, and unmarried sisters. In an effort to bring to a conclusion his attempts to obtain the chair at Padua, he decided to go to Venice and personally offer his services to the Venetian Republic. The good impression he made on prominent Venetians and the help of influential friends won him the desired position. In the fall of 1592, having received the Grand Duke's permission to leave Tuscany and his lectureship in Pisa, he settled in Padua.

3. Good Times

GALILEO WAS 28 years old when he went to Padua. There, says one of his earliest biographers, "he got himself a small house for a home, not far from the most famous church [and monastery] of Santa Giustina. The nearness of this place was of great convenience to him, since the abbot then heading the monastery was a gentleman from Verona of courteous manners, and an appreciator, in no small degree, of mathematics. This fact gave [Galileo] a chance to enter into friendship with [the abbot] and as a consequence he was provided with a few necessary utensils and pieces of furniture, like beds, chairs and similar things, for which he had no little need. . . ."

We learn from this description that, upon his ar-

rival in Padua, Galileo was still poor and in need of
help. Despite this difficult start, he spent the "best eighteen years of his life" in Padua. They were his best not
only for his scientific achievements but also for other
reasons. He had no great worries yet and could experience the full satisfaction of the successful researcher
whose work is always in harmony with the needs of
his mind. He made many congenial friends, because his
jovial temperament and witty conversation attracted all
intelligent persons with whom he came in contact. Last
but not least, he had the privilege of living in the Republic of Venice and close enough to that city, even
in those days of horses and carriages, to enjoy its attractions.

The *Serenissima* (most serene) Republic of Venice
was the most enlightened and liberal state of that period.
It was governed by members of some five hundred noble
families who, generation after generation, gave the republic statesmen and great military leaders. The Doge,
the head of the republic, was chosen from among them,
as were the two bodies controlling his authority, the
Council of Ten and the Senate. Thus the governing
nobility formed a closed caste, a system that would not
receive approval in our democratic age. At that time,
however, most countries were under absolute rulers,
often not as wise, well educated, and fair to the people
as the Venetian noblemen.

Centuries of trade with the Levant (as the Near East
was called) and repeated victories over the Turks had
brought great wealth to the republic. At the end of the
sixteenth century it was still one of the richest states in
Europe, although it showed some signs of decadence.
The city of Venice was then, as it is now, a unique city
with a unique charm. It was built on water, along innumerable canals and the open shores of a lagoon. Two
civilizations had joined to give it splendor. Western
artists had built its palaces and churches, and they had
poured over these the richest ornaments from the Levant:
colored marbles and porphyries, mosaics glittering with

gold, statues, sculptures, and paintings of strange symbols and fantastic animals.

A VENETIAN NOBLEMAN

Galileo made friends among the nobles and the rich Venetians, most of whom were extremely cultivated and well versed in all branches of learning. Several of them held gatherings to discuss scientific matters, and a few even had laboratories of sorts or shops in their homes. At a time when scientific meetings were not yet in fashion and universities had no laboratories, these private intellectual activities were very important.

Of Galileo's Venetian friends, none was so dear to him, so faithful over so many years, as Giovanni Francesco Sagredo. Sagredo, a few years younger than Galileo, belonged to one of the noble Venetian families forming the ruling class. At 25, in 1596, he became a member of the Supreme Council of the republic and devoted himself to the cares of government. His personal interests, however, leaned toward letters and science, and he had friends among writers and philosophers. He felt admiration and affection for Galileo, and more than once he used his influence to improve the scientist's position at the university. He was always willing to give practical advice, to share fun, jokes, and escapades. Of this fun and these escapades there is many a nostalgic mention in Sagredo's later correspondence with Galileo.

Galileo reciprocated this friendship, and several years after Sagredo's premature death he immortalized his young friend as a character in his two best books, *Dialogue of the Two Greatest Systems of the World* and *Two New Sciences*. In these, Sagredo emerges as a typical example of a versatile, cultivated man.

The enlightenment of the Venetians was reflected in the policies of the University of Padua, where teachers, the best from all Europe, had complete intellectual freedom. Freedom is an intangible asset which we appreciate at its full value only if we are deprived of it. Galileo was to learn this through bitter experience after he left

Padua. Then he realized how essential it is for a
scientist to live and work where there is no outside
interference with the development of scientific thought.

THE SCHOOL OF MEDICINE

By the time Galileo joined its faculty, the ancient University of Padua (founded in 1222) was flourishing and
its fame spreading over Europe. Many of the professors were still conservative philosophers who passively
accepted Aristotle's authority. But there were also new
trends much better suited to Galileo's independent
spirit. It was certainly important in the development of
his philosophy of science that there existed in Padua a
school of medicine with a strong experimental bent. The
school had been established in the thirteenth century.
Only 30 years before Galileo went to Padua, the
Belgian Andreas Vesalius, professor of anatomy, became
famous for performing his own dissections instead of
relying on assistants. In Galileo's time another great
biologist had Vesalius's position, Hieronymus Fabricius
of Aquapendente. Fabricius, an anatomist, physiologist,
embryologist, and Galileo's physician, directed the construction in Padua of the first anatomical theater ever
built. Among his students, the most famous was William Harvey, the founder of modern physiology, who received his degree in Padua in 1602.

Of the professional relations between Galileo and
Fabricius, little is known. Galileo had great confidence
in his physician, whose methods of diagnosing and
treating diseases he greatly praised and whose pills he
kept at hand. After an illness he would stay in bed
and wait for Fabricius's permission to get up, more pa-
tiently than we would expect of a man of Galileo's
temperament. It seems unlikely that in their intercourse
the biologist and the natural philosopher should not
have discussed scientific questions and influenced each
other's thought, perhaps even affecting the course of
physics and biology. There is no evidence, for instance,
that Fabricius's pupil, William Harvey, ever met Galileo;

but Harvey's method of limiting research in biology to problems that could be solved experimentally was so similar to Galileo's that it suggests strong ties between physics and biology in the Padua of Galileo's times.

Another of Galileo's colleagues in Padua, the physician Sanctorius, was introducing measurement in medicine. It is interesting to note that Sanctorius used and described the "pulsilogium," an instrument to measure the rate of pulse beat, and therefore the fever of his patients. This instrument was based on the properties of pendulums discovered by Galileo, and it was first built by Galileo, according to Viviani. Sanctorius was the first physician to use a thermometer, whose invention Galileo claimed for himself. On this point there is considerable debate, and the thermometer seems to have been invented independently by several people at different places.

PUBLIC AND PRIVATE TEACHING

In this climate, where Aristotelianism and modern currents went side by side, Galileo started his university teaching. We must believe that his classes were well attended from the very beginning; his formal opening lecture drew this comment from a learned Danish man: *"Exordium erat splendidum, in magna auditorum frequentia* [The opening was splendid, with a great attendance of hearers]."

Students came to Padua in large numbers, from all countries and all walks of life, and formed a colorful crowd. Many wore rich clothes of brocades, silks, and velvets and often carried glittering swords. The rich and the noble students brought along their private tutors, their secretaries, and their servants—some, we are told, sent their servants to school while they did the town.

Confronted with students of many nationalities, Galileo, like the other professors, lectured in Latin. The nucleus of his audience was the medical student who took mathematics in order to understand cosmography and astrology; they needed astrology, for despite the new trends in medicine all good physicians were expected to draw horoscopes.

Galileo was an exceptional teacher. From known and common experience he took his students by easy steps to abstruse heights. He presented prevailing opinions which he thought wrong as if they were right, and when he had convinced his audience, he tore these same opinions apart bit by bit, showing where they failed. His classes were popular and he had to move to larger and larger classrooms. Many travelers stopped in Padua to hear at least one of his lectures.

To augment his salary, which was rather meager at first, Galileo followed the current custom of giving private lessons. His private pupils were not only the young university students but also, as time went on and his reputation grew, foreign noblemen in search of higher learning. Several of his private pupils were boarders in his home. The most serious students in Padua were anxious to live with the best teachers, to enter into familiar relations with them and thus receive informal tutoring. To this incentive Galileo added that of an easy hospitality, enlivened by his affable manners and original conversation and perhaps also by the excellent food and wine that he is said to have served always, in the good Tuscan tradition. He had moved to a larger house where he could entertain up to 20 paying guests at a time. Gradually, as these left and returned to their homes, his influence as a scientific thinker spread in Italy and abroad.

For use both in his university classes and private teaching Galileo wrote several works, which he circulated in manuscript copies, some of which have been lost. Those preserved show that Galileo taught not only fundamental science, like cosmography, but also applied science, directed at solving technical and practical problems. In fact, he wrote treatises on *Fortifications, Military Constructions,* and *Mechanics.* This last, the most important, is a true engineering textbook, in which Galileo described the action of simple machines, like the lever, pulley, screw, and inclined plane. He also dealt with problems of falling bodies and resistance of materials, problems which were dear to him and which he took up again in later years.

GALILEO'S MILITARY COMPASS

In 1597 Galileo built an instrument which illustrates both his competence in mathematical calculation and his great craftsmanship: the "geometric and military compass," now called "sector." We may consider it a combination proportional divider and slide rule. It bore many scales and helped to solve a large number of mathematical and geometrical problems (including the extraction of square and cubic roots). At a time when logarithms had not yet been invented, his compass was very useful. Because the end of the sixteenth century was a period of conflict, of civil and international wars, Galileo stressed the military uses of this instrument—for instance, to determine relations of weight and size of cannon balls, to regulate the front and side formations of armies, to measure the inclination of a scarp or wall.

Galileo's compass sold so well that it became a good source of income. He could not keep up with demand. To produce it in quantity, he hired a craftsman who came to live in his home. Thus Galileo started his own shop, a project of no little importance, considering that there was no physics laboratory at the university and that Galileo needed accurate instruments and equipment for his experiments.

It should not surprise us that the compass sold so well, although it was not one of Galileo's greatest achievements. Then, as now, few people were able to appreciate purely scientific work. The great majority judged the products of science either for their practical value or, on emotional grounds, as factors that might subvert established beliefs and usages. Most people set a higher value on a useful technical invention, even if of modest import, than on a great scientific discovery.

We have mentioned Galileo's compass mainly because it provides further evidence of his complex personality. From all he did in the early part of his stay in Padua, Galileo emerges as an unusual example of the well-rounded man in whom intelligence and craftsmanship, speculative power and practical sense were harmoniously fused. He could easily pass from teaching to manual work to purely scientific meditation.

Galileo's most important scientific achievements in the period at Padua were in two fields: the study of motion and astronomy. To his astronomical work, through which he was dragged into the controversy with the Church and the Aristotelians, we shall devote the next chapter.

In the field of motion he resumed the study started in Pisa with observations on pendulums and falling bodies and developed it to such an extent that he could call it "a science entirely new, for no one else, either ancient or modern, has discovered any of the most admirable principles that I prove to be in natural and violent motion. . . ." (Aristotle taught that everything had its "natural" place and "natural" motion. The "natural" motion of smoke was upward, of a stone, downward. Deflected smoke or a thrown stone acquired "violent" motion.)

Galileo did not publish his observations on motion until 1638, when they appeared in *Two New Sciences*. Only through his scientific correspondence do we know that he had laid the foundations for this book in Padua and can we follow some of the steps of that long process of trial and error which eventually made him the first modern physicist.

In the intervening years Galileo never ceased to review and refine his conclusions or to work on unanswered questions. So *Two New Sciences* is an end product, the product of mature thought. Because it is not always possible to distinguish between earlier and later work in this book, to know exactly what he did in Padua and what afterward, we shall follow Galileo's example and defer discussion of his work in mechanics to a later chapter.

MAGNETS

Instead, we shall consider another instance of his lively
interest in everything new both in science and in tech-
nology; namely, his work on magnetism and on natural
magnets. In 1600 William Gilbert of Colchester, English
physician at the court of Queen Elizabeth, published a
book, *De Magnete* (*Concerning the Magnet*). Magnetism,
the property of loadstone and other substances to attract
iron, was then well known, and navigators were using
magnetic needles to direct the course of their ships. Gilbert summarized current knowledge of magnetism and
described his own experiments on magnetic and electrical
attractions. What is more important, he advanced the
hypothesis that the earth is nothing but a huge magnet,
which fact, he said, explains both the declination and the
dip of the magnetic needle. (Magnetic needles orient approximately from north to south, but not exactly. The
angle they form with the true north-south direction is
called *declination;* the angle with the horizontal is called
inclination or *dip.*)

At once Galileo conceived a great admiration for Gilbert, as he did for most truly great people. In the *Dialogue* he said of Gilbert: "I highly praise, admire and
envy this author because such a stupendous concept came
into his mind about a thing which innumerable men of
sublime intellect handled and none understood."

Galileo, who did not accept the authority of the written word, repeated Gilbert's experiments and performed
others. His vivid, almost childlike excitement over magnetic properties is evident in many of his letters of that
period. He shopped around among "antiquarians" and
"lapidaries," seeking more and more powerful loadstones;
he wrote to his friends both in Florence and Venice to
make them share the wonders of magnets. He even became
middleman in the purchase of an especially good loadstone for the Grand Duke of Tuscany and then spent
much time improving it. Gilbert had found that if he put
a thin iron jacket, which he called armature (or "body

armor"), around a magnet, the apparatus could support a much greater weight than the magnet alone. With his skilled hands Galileo improved armatures to the point that a small loadstone weighing six ounces supported a weight of 160 ounces.

The action of armatures, however, remained an unsolved problem which puzzled Galileo for many years, and to explain it he resumed his experiments on magnetism in the later part of his life. In the *Dialogue* he tells us that he became convinced that through armatures "the power and force of the stone is not increased . . . there is no change in the force, but merely something new in its effect. *And since for a new effect there must be a new cause,* we seek what is newly introduced in the act of supporting the iron via the armature, and no other change is found than a difference in contact. For where iron originally touched loadstone, now iron touches iron, and it is necessary *that the differences in these contacts cause the differences in the results.*" [Editors' italics.]

He then asked himself what could cause this difference in contact and made the hypothesis that the loadstone was porous and contained impurities; these prevented full contact between the loadstone and the supported weight. To check whether this was true, he had the surface of a loadstone polished and smoothed, and to his satisfaction he could verify that impurities showed up as spots on the stone's surface and that iron filings brought near it leaped to the stone but not to the spots. The thin iron of the armature became uniformly magnetized and thus improved the contact with the suspended weight.

Galileo's research on magnetism is a minor episode in his scientific activity, not very significant as a scientific contribution. But it is worth noticing that here as elsewhere Galileo gained understanding of a phenomenon by searching for the cause of an effect. The need for this search seems obvious to us, but one of Galileo's great contributions to science consists precisely in his having shed light on this point. Continually he stressed that "for each effect there is a prime and true cause."

"Between cause and effect," he wrote, "there exists a firm and constant relation, such that necessarily each time that we observe a variation in the effect there must be a firm and constant alteration in the cause." Science, he taught, consists in great part in finding simple relations between causes and effects in apparently complicated phenomena.

A few more words will complete the general picture of Galileo's years in Padua. He enjoyed not only science but also music, art, and literature. He left notes on two great Italian heroic poems, the *Gerusalemme Liberata* by Torquato Tasso, which had just been published, and the older *Orlando Furioso* by Ludovico Ariosto. From these notes, probably written in Padua, we learn that in a literary work Galileo sought the same clarity and coherence that he deemed necessary to science. He disliked Tasso's prolix poem, its artificiality, its improbable situations and poorly defined characters. By contrast, he considered Ariosto a "magnificent, rich and admirable" poet. He often quoted octaves from *Orlando Furioso* and said that he had learned clarity and logical expression from it. He was used to saying that "between the one and the other [poets] he felt the same difference that eating cucumbers after relishing savory cantaloupes brought to his taste, or palate."

Galileo often went to Venice, which at that time offered varied entertainment: good concerts and the theater; lively dances; great religious celebrations; and spectacular festivals, when men and women in opulent costumes gathered in the Piazza San Marco, the largest square, and old Venetian galleons were anchored at the docks.

Of great men's lighter sides there is usually little record, as we may expect, for only their important work survives. We have a few hints of Galileo's light side from his correspondence. We have said that his friend Sagredo, in his nostalgic reminiscences of later years, recalled practical jokes and escapades shared with Galileo and Galileo's witty conversation. Other friends and pupils recalled the good times they had drinking wine with their teacher or friend, and some of their letters accompanied the shipment of a barrel of wine or expressed thanks for receiving one.

AN UNFORTUNATE TRIP

Galileo's great love of the open air took him on frequent trips to the country, and one of these resulted in a freak accident that impaired his health for the rest of his life. He and two of his friends were guests in a villa on the hills near Padua, according to Viviani, who tells this episode and is confirmed by Galileo's son. The villa, like a few others in that neighborhood, had a sort of primitive air-conditioning system. Through underground ducts the villa was connected to natural caves where the air stayed cool and moist through the hottest summer, and this air flowed into the house through an opening that was closed off when the house was not too warm. Galileo visited the villa on a very hot day, and in the noon hour, probably after too large a meal, he and his two friends went to sleep near the opening that let in the cool, humid air. When they awoke, the three men felt ill and racked with pains. The illness was so severe that one of the friends died. Galileo recovered, but from then on he was frequently afflicted with severe attacks of arthritic pains which often kept him in bed, sometimes for long periods.

During his stay in Padua, Galileo met a young woman, Marina Gamba, and fell in love with her. They had three children, although they were never married. In those times few frowned upon such relationships. Virginia, Livia, and Vincenzio were born, respectively, in 1600, 1601, and 1606. Virginia, the eldest daughter, became Galileo's favorite and was a source of great comfort to him in his old age. When she was born, Galileo followed current custom and wrote her horoscope in Latin. Like most horoscopes, it was vague and not too meaningful, and its predictions did not come true.

Galileo's responsibilities were increasing, since in leaving Tuscany he had not cut his ties with his family, nor had he given up his duties toward them. On the contrary, he received frequent visits in Padua from his mother, sisters, and brother; he gave them moral and financial support and helped them in every way.

In view of subsequent events, it is significant that he spent most of his summer vacations in Florence and that from the summer of 1605 he was private tutor of mathematics to young Prince Cosimo de' Medici, the Grand Duke's son. Teacher and pupil became sincerely attached to each other by mutual affection and deference. Galileo, who had been born a subject of the Medicis, felt this allegiance deeply to the end of his life.

4. The Telescope

ON OCTOBER 9, 1604, a new star appeared in the sky.

New stars appear from time to time. We now know that they are stars of low luminosity which suddenly brighten up because the gases of which they are made explode; they then slowly return to their initial luminosity. We call them "novae" or "supernovae," according to their brightness. The explosions producing new stars are similar to those achieved on earth with hydrogen bombs; they occur by the fusion of light elements, like hydrogen, into heavier elements with gigantic release of energy.

Only one other new star had been observed in modern times: in 1572, when Galileo was a small boy. Because of the rarity of the phenomenon and the fact that everybody then related human events to celestial phenomena, the new star of 1604 created general excitement. It especially aroused Galileo's scientific curiosity, and he undertook to observe it at once. This, his first astronomical observation, initiated his long career as stargazer. Yet his interest in astronomy and his theoretical studies in this field had begun much earlier. To trace their origin we must jump back in history to 21 years before he was born.

COPERNICUS

It was the year 1543. A book had been published in Germany, and its author, the Polish astronomer and Catholic canon Nicolaus Copernicus, received the first copy on his deathbed, then died the same day. His book, *De Revolutionibus,* was to change the way of thinking of the Western world, but not at once.

In *De Revolutionibus,* Copernicus analyzed the two systems of the universe generally accepted in his time: the system described by Aristotle, in which the sun, moon, planets, and stars moved in concentric circles around the earth, and the system described by Ptolemy of Alexandria in the *Almagest.**

Ptolemy, who lived in the second century, was a great astronomer, the last of the great Greek astronomers. He used a scientific instrument called the astrolabe to make his observations, and he applied mathematics to astronomy. In the *Almagest* he gathered all the astronomical knowledge of his times and added his own contributions. Like Aristotle, Ptolemy believed that the earth stood still at the center of the universe, while the planets and sun turned around it. In order to explain the apparent motions of the planets, he used a system of spheres and circles whose centers moved along other circles. For 14 centuries, until Copernicus's time, astronomers had accepted Ptolemy's system. Aristotelian philosophers had considered its geometrical structure an abstraction, an expedient to describe motion, which did not contradict Aristotle's basic ideas on the universe.

---

* Ptolemy's works, like those of Aristotle, were saved by the Arabs. It is interesting to note that the name Almagest comes from the Arab *al* (the) and the Greek *megiste* (greatest), and so it means "the greatest book." The Greek name of Ptolemy's book was *Syntaxis.*

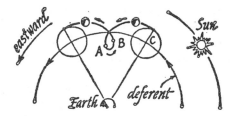

Motion of planets. By varying the radii of the epicycle and deferent, and the speeds of the deferent and planet on the epicycle, the Greeks could find the type of path that best fitted the apparent motion of each planet.

After analyzing Aristotle's and Ptolemy's theories, Copernicus presented a system in which the earth was not at rest but moved in an orbit around the sun, as did the other planets. The moon revolved around the earth while the earth moved. The stars did not revolve, as in the earth-centered systems of Aristotle and Ptolemy, but were fixed; their apparent motion was attributed to the daily rotation of the earth on its axis.

An accurate description of planetary motion would have been of truly practical importance in Copernicus's time. It was the time of great sea voyages and explorations, and one of the requirements of navigators for directing the course of their ships was to know the precise positions of planets at successive times. But it was not possible to predict these positions when the motion of planets was not fully understood. Further, the Church of Rome also needed precise astronomical data in order to reform the calendar, which had become quite inaccurate.

Yet Copernicus hesitated a long time before publishing his book. He knew that the idea of a moving earth went against a tradition that had lasted 14 centuries, against the theologians' literal interpretation of some passages of the Bible, and against the accepted view that man, living at the center of the universe, occupied a privileged position in God's creation.

When *De Revolutionibus* eventually was published, most

people found it too abstruse, and few read it. But as time went on, it attracted more scientists. Some who were able to understand the book and who did not reject the assumption of a moving earth lectured on it. A few of them traveled and lectured all over Europe, including Italy, which was then the most learned country.

Some 50 years after Copernicus's death his followers still were few. Among them were Galileo and the young German astronomer Johannes Kepler (1571-1630). We mention them together because it is from letters they wrote to each other in 1597 that we know they were both Copernicans.

KEPLER

Kepler was to become one of the greatest astronomers, perhaps the greatest of all time. He was to discover the three laws, now bearing his name, that are the true description of the planets' motions *around* the sun. These laws state, in essence, that planets move not in circular orbits but in ellipses, in one of the foci of which is the sun; they move more rapidly when they are closer to the sun than when farther away from it; the time that each planet takes to describe its orbit depends on its distance from the sun and increases with it according to a definite proportion. Copernicus, in order to account for the observed irregularities in the apparent motion of the moon and planets, had devised a system of circles moving on circles which was almost as complicated as Ptolemy's. Kepler's laws not only described the true motion but revealed the solar system's essential simplicity and beauty. A simple formulation often is a necessary step to further progress in science.

Kepler, however, had not yet gone this far by 1597, the year that interests us at this point. He was then only 26 years old and had just published a book, *Mysterium Cosmographicum,* in which he presented some proofs in favor of the Copernican system. His views were not yet as firm and definite as they later became, and, perhaps to get the reaction of another scientist, he sent a copy of his book to Galileo, whose fame was rapidly spreading.

Galileo read the preface of the *Mysterium* and was so impressed that he wrote to Kepler at once, congratulating him on his discoveries. From what Galileo then wrote we learn that he had been a follower of Copernicus long before 1597, but a rather timid one. In his letter he promised that he would soon read the rest of Kepler's book and then went on: "This I shall do the more willingly because many years ago I became a convert to the opinions of Copernicus. . . . I have arranged many arguments and confutations of the opposite opinions, which, however, I have not yet dared to publish, fearing the fate of our master Copernicus, who has earned immortal fame among few, yet by an infinite number (for such is the number of fools) is held in contempt and derision."

Kepler replied, expressing joy for the newly acquired friendship and for the fact that Galileo shared his own opinions. He also encouraged Galileo to persevere on the road of truth even though he were to be "among few." *"Confide, Galileae, et progredere,"* he said (their correspondence was, of course, in Latin). "Have faith, Galileo, and go on."

Despite Kepler's enthusiasm, Galileo kept his views to himself and his friends for several years more, and when he gave an astronomy course at the university, he respected tradition and taught the Ptolemaic system. In reality he did not pay much attention to Kepler's work, and it is almost certain that he never read past the preface of the *Mysterium*. Kepler's Latin was difficult, often obscure, and his way of thinking was much more fanciful and mystical than Galileo's.

If a book could not arouse Galileo, a new star could. The sudden appearance of a new star in that sky which official science considered unchangeable was to him a stronger incentive than any written word. The new star of 1604 was to change the course of his life.

Shortly after it appeared, he started to make systematic observations, to measure its height in the sky, to check whether its position in respect to other stars varied from night to night or from hour to hour of the same night. After having watched it for several weeks, he gave three public lectures to describe his findings and drew large crowds.

His interpretation of the nature of the new star was wrong and naïve. He thought that it might be made of very tenuous terrestrial exhalations reflecting the light of the sun. He was right, however, in stating that the new star was much farther away from the earth than the moon and the planets, and was located among the fixed stars. He had come to this conclusion from the fact that the position of the star did not vary in respect to fixed stars, or, as the astronomers say, it did not have parallax. (Parallax in astronomy is the difference in direction of a heavenly body when it is observed from two points on the earth's surface. A similar change in direction may be seen also when looking at a closer object: if one holds a finger in front of his face and looks at it with one eye closed, then with the other eye closed, the finger will seem to have moved.)

The Peripatetics,* as the followers of Aristotle were called, could not accept the view that the new star was located among the fixed stars. They believed that the celestial region was perfect and immutable. They said that since the star was new it could not be in the immutable celestial region, so it *must* be in the terrestrial region, closer to the earth than the moon, and Galileo *must* be wrong.

This was the kind of reasoning that Galileo opposed and that aroused his fighting spirit. Thus he was drawn into a controversy in which he fought against the Aristotelians and for the Copernicans for the rest of his life. Perhaps this controversy, if limited to the new star, would not have been sufficient to originate the drama in which Galileo was actor and victim. But something happened.

---

* Aristotle used to teach and talk to his pupils while walking. Hence the name "Peripatetics," meaning "strollers."

NEWS OF A TELESCOPE

In the early summer of 1609 he was in Venice when he heard that "a certain Fleming [Hans Lippershey] had constructed an eye-glass by means of which visible objects, though very distant from the eye, were distinctly seen as if nearby." Galileo went back to Padua at once and thought how he could build a similar instrument. His reasoning was rather simple: "The device needs either a single glass or more than one. It cannot consist of one glass because the shape of this would have to be either convex (that is, thicker in the middle than at the edges), or concave (that is, thinner in the middle) or bounded by parallel surfaces. But this last does not alter visible objects at all, either by enlarging or by making them smaller; the concave diminishes them; and the convex enlarges them well but makes them indistinct and blurred. So one glass is not sufficient. Passing then to two, and knowing that a glass with parallel surfaces alters nothing, as already said, I concluded that the effect would not be achieved by combining this glass with either of the other two. So I was reduced to trying the combination of the other two, namely a convex and a concave glass, and seeing what it would do. I found that this gave me what I sought." In other words, he found that the system of two lenses, when these were placed at a suitable distance, gave a *distinct* and much enlarged view of faraway objects.

Galileo mounted the two lenses at the ends of a lead tube, and so his first telescope came into being. For some time Galileo called his instrument simply *occhiale*, which means "eyeglass," or by the Latin word *perspicillum,* and only later did he use the word "telescope."

At once he realized the great importance of his instrument, both in science and in practical. life. He realized, for instance, that it might substantially modify the art of war, that it would permit sighting "the enemy at much greater distance than usually . . . discover him more than two hours before he can see us . . . judge his forces, so

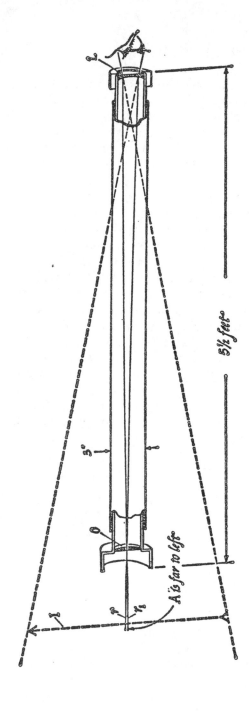

Galileo's Telescope. This diagram indicates how the telescope magnifies a distant object, *A.* Light rays from the object (represented here only by the pair *r, r₁*) go through the converging lens (*O*) and the eyepiece (*L*) at the other end of the tube. This diverging lens widens the angle of the narrow beam so that to the eye near the lens the rays appear to come from the directions shown by the dashed lines. All rays from the object seem to come from the area *I*, forming an image which appears much larger and closer than the object itself does to the naked eye.

as to get ready to give him chase or battle, or to retreat. . . ."

Galileo must have shown his instrument to friends or
talked about it, for soon the rumor spread that he had
invented it. In a letter to his brother-in-law, Landucci,
dated August 29, 1609, he related the events that ensued:
"And news having reached Venice . . . I was called by
the Most Serene Signory six days ago, to whom it was my
duty to show it as well as to the whole Senate, to the
intense astonishment of all."

GALILEO PRESENTS HIS TELESCOPE

It is easy to reconstruct the scene in the Palazzo Ducale,
the grandiose palace of the government. Senators and
noblemen, who in contemporary paintings are represented
with flowing beards and clad in silken togas and small
pillbox caps, were seated on a platform. The walls and
ceilings of the magnificent halls were adorned—as they
still are—with the masterpieces of Titian and Tintoretto.
The professor from Padua, Galileo, also in toga, stood
with his telescope on the marble floor. "Galileo's gun,"
in the words of Antonio Prioli, a university administrator
who was there that day, "was of sheet metal, covered on
the outside with crimson sateen, of about the length of
24 inches and the breadth of a *scudo* (a silver coin, approximately 1¾ inches in diameter) with two glasses
[one] at each end, one concave and the other not."

The immense hall was not large enough; Galileo needed
more space to demonstrate the power of his "gun." With
many of those present he went outside, down the monumental stairs of the Palazzo Ducale, and up the steeper
stairs of the Campanile, the bell tower of San Marco
across the Piazzetta. "And very many were the patricians
and senators who, although aged, have more than once
climbed the stairs of the highest campanili of Venice, to
detect sails and vessels on the sea, so far away that
coming under full sail toward the harbor, two hours or
more passed before they could be seen without my eye-

glass: because in fact the effect of this instrument is to represent an object that is, for example, 50 miles off, as large and near as if it were only five miles away.

"Now . . . seeing that this Most Serene Prince desired it, I resolved on the 25th to appear before the *Collegio* [the high advisory body to the Doge] and to make a free gift of it to His Serenity. And I having been ordered to go out of the Collegio and wait . . . shortly afterwards the Most Excellent Procurator Prioli, one of the Administrators of the University, came out of the Collegio. He took me by the hand and said that the Most Excellent Collegio . . . had immediately ordered the Most Illustrious Administrators that, with my assent, they were to renew my appointment for life with an annual stipend of 1,000 florins. . . . I, knowing that the wings of hope are dilatory and those of fortune swift, said that I was content with whatever pleased His Serenity."

5. The Universe Through the Telescope

ONCE MORE GALILEO showed his craftsmanship. The first telescope he built had a magnification of three diameters, the second of eight diameters, and finally he built one magnifying 33 diameters. He increased the size of his lenses but stopped at the point where further increase would cause distortion of images. Later he built many more telescopes, some of which are preserved and are still objects of admiration.

Yet, more than a clever craftsman, he was a true scientist. No one before him so clearly realized that instruments serve two essential purposes: First, they make observations both objective and quantitative, transforming them into "measures." Second, they extend the power of human senses beyond their normal physiological limits. No instrument but the telescope could give Galileo the feeling of how much, above all expectations, the power of the senses could be enlarged. For the first time eyes could see very distant things as well as if they were close by and discover the existence of objects that up to then had been too far away to be visible.

To make full use of the telescope as a scientific instrument Galileo did something that no one had done before. He turned it to the sky.

We are so used to the "extension of our senses"—to loud-speakers and other amplifiers increasing the power of our hearing; to powerful telescopes and microscopes enlarging our vision—that we cannot easily recapture the intensity of Galileo's experience. He, first of all men, could realize how much larger God's creation is than that which is revealed to the unaided senses, and for this privilege he devoutly offered frequent thanks to God in his writings.

For tens of centuries men had seen the same objects in the sky, and they had come to believe that they had learned all there was to be learned in it. But Galileo saw new depths and a new population in the heavens. Everywhere he turned his telescope he saw stars never seen before, in the most crowded constellations and in the thinly scattered regions. He recognized that the Milky Way is not a long luminous cloud but is made of numberless stars so dim and close together that the naked eye cannot distinguish them.

NEW ASPECTS OF THE MOON

The moon appeared very different to him, and among its new strange features were prominences and cavities in its surface. ". . . Many of the prominences there are in all respects similar to our most rugged and steepest mountains, and among them one can see uninterrupted stretches hundreds of miles long. Others are in more compact groups, and there are also many isolated and solitary peaks, precipitous and craggy. But most frequent there are certain ridges . . . very much raised, which surround and enclose plains of different sizes and shapes, but mostly circular. In the middle of many of these there is a very high mountain, and a few are filled with rather dark matter. . . ."

Galileo calculated the height of lunar mountains and concluded correctly that some are as high as four miles. He thought that this was higher than *any* mountain on the earth. (Mount Everest, we know, is about five and a half miles high.)

JUPITER'S SATELLITES

The telescope held still more marvels in store. Galileo soon discovered four small planets (or satellites) revolving around the planet Jupiter, as the moon revolves around the earth. (More satellites of Jupiter were discovered later, with more powerful telescopes than his and with photographic methods. There are twelve in all.) Galileo attached a great importance to this astronomical discovery, and even in later years it remained his favorite. Accordingly he named Jupiter's satellites *Medicean Planets,* in honor of his beloved pupil Cosimo, who had become Grand Duke, and in honor of Cosimo's family.

"Behold, then, four stars reserved to bear your famous name," Galileo wrote in the dedication of his book *The Starry Messenger,* "bodies which belong not to the inconspicuous multitude of fixed stars, but the bright ranks of the planets. Variously moving about most noble Jupiter as children of his own, they complete their orbits with marvelous velocity—at the same time executing with one harmonious accord mighty revolutions every dozen years about the center of the universe; that is, the sun."

We see from this paragraph that Galileo not only had observed Jupiter's planets but also was studying their motion. Indeed, he hoped to arrive at a complete understanding of this motion and to predict the satellites' positions so accurately that navigators would then be able to rely on them to direct the course of their ships. Jupiter's satellites are more suitable than planets for this purpose

duos proxime menses observationes circa eorundem lationes ac mu-
tationes habitas, aperiamus ac promulgemus; Astronomos omnes con-
vocantes, ut ad illorum periodos inquirendas atque definiendas se con-
ferant, quod nobis in hanc usque diem, ob temporis angustiam, assequi
minime licuit. Illos tamen iterum monitos facimus, ne ad talem inspectionem incassum accedant, Perspicillo exactissimo opus esse, et
quale in principio sermonis huius descripsimus.

Die itaque septima Ianuarii, instantis anni millesimi sexcentesimi
decimi, hora sequentis noctis prima, cum caelestia sidera per Perspicillum spectarem, Iuppiter sese obviam fecit; cumque admodum 10
excellens mihi parassem instrumentum (quod antea ob alterius or-
gani debilitatem minime contigerat), tres illi adstare Stellulas, exiguas
quidem, veruntamen clarissimas, cognovi; quae, licet e numero iner-
rantium a me crederentur, nonnullam tamen intulerunt admiratio-
nem, eo quod secundum exactam lineam rectam atque Eclipticae
parallelam dispositae videbantur, ac caeteris magnitudine paribus
splendidiores. Eratque illarum inter se et ad Iovem talis constitutio :

Ori.   *   *   ◯   *   Occ.

ex parte scilicet orientali duae aderant Stellae, una vero occasum ver-
sus. Orientalior atque occidentalis, reliqua paulo maiores apparebant: 20
de distantia inter ipsas et Iovem minime sollicitus fui; fixae enim, uti
diximus primo, creditae fuerunt. Cum autem die octava, nescio quo fato
ductus, ad inspectionem eandem reversus essem, longe aliam consti-
tutionem reperi: erant enim tres Stellulae occidentales omnes, a Iove
atque inter se, quam superiori nocte, viciniores, paribusque interstitiis
mutuo disseparatae, veluti apposita praesefert delineatio.

Ori.   ◯   *   *   *   Occ.

Hic, licet ad mutuam Stellarum appropinquationem minime cogita-
tionem appulissem, haesitare tamen coepi, quonam pacto Iuppiter ab
omnibus praedictis fixis posset orientalior reperiri, cum a binis ex 30
illis pridie occidentalis fuisset : ac proinde veritus sum ne forte secus
a computo astronomico directus foret, ac propterea motu proprio
Stellas illas antevertisset. Quapropter maximo cum desiderio sequen-
tem expectavi noctem; verum a spe frustratus fui, nubibus enim uu-
diquaque obductum fuit caelum.

Discovery of Jupiter's Moons. A page from *The Starry Messenger,* in which Galileo describes some positions of Jupiter's
satellites.

because their periods of revolution are shorter. (The satellite Io, 262,000 miles from Jupiter, completes an orbit around Jupiter in 42½ hours.) He spent evening after evening, year after year, studying the motion of the Medicean Planets with his telescope, indifferent to the chill of night that increased his arthritic pains or to the strain that prolonged observation put on his eyes. He came close to his goal but did not quite reach it.

When he turned his telescope on the planet Saturn, he found that this did not always look like a round body but seemed of strange variable shape. He thought Saturn to be "three-bodied; that is, it . . . was an aggregate of three stars arranged in a straight line parallel to the ecliptic, the central star being much larger than the others." His telescope was not sufficiently powerful to let him distinguish the three, possibly four, rings we now know are around Saturn. It was the Dutch astronomer and inventor Christian Huygens (1629-95) who discovered Saturn's rings.

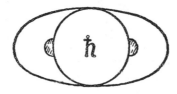

Galileo's Sketch of Saturn. The planet's rings could not be distinguished with his telescope.

After first observing Saturn, Galileo "announced" his observation in a Latin anagram: *Salve umbistineum geminatum Martia proles*, which he explained shortly afterward: *Altissimum planetam tergeminum observavi.* [I observed the highest planet (to be) three-bodied.]

THE PHASES OF VENUS

Less than a month later Galileo announced "another detail newly observed by me, which drags along the decision of very great controversies in astronomy, and in particular contains a vigorous argument for the Pythagorean and Copernican constitution."

Although Galileo called it "a detail," this was a true discovery, the discovery of the phases of Venus. On this occasion, as after his observation of Saturn's shape, Galileo wrote an anagram: *Haec immatura a me frustra leguntur o y,* which he later explained as meaning: *Cynthiae figuras aemulatur mater amorum* [The mother of love emulates the shapes of Cynthia (the moon)]. He described the phases of Venus toward the end of 1610:

". . . about three months ago I began to observe Venus with my instrument, and I saw it of rounded shape and very small: It went on growing daily in size, still keeping its roundness, until eventually, arriving at a very great distance from the Sun, it started to lose its roundness on its eastern side, and in a few days diminished to a half circle. In this shape it stayed several days, growing however in size: now it starts to become of horned shape, and so long as it will appear as evening star, its small horns will diminish [it will become thinner] until it will vanish; but when it will appear again as morning star, it will be seen with very thin horns, and also turned against the Sun, and it will grow to become semicircular at its maximum departure [from the sun]. It will then remain semicircular for several days, decreasing however in size; and then from semicircular it will get to full roundness in a few days, and after that for several months it will be seen completely round, both when [it appears] as morning star and evening star, but rather small in size."

Copernicus had been disappointed not to see phases of Venus and variations in the apparent size of the planet. The fact is that the unaided eye cannot detect them. Their discovery through the telescope came as a strong argument against the hypothesis of an earth-centered universe —a much stronger argument than the Medicean Planets provided. (Indeed, to the foes of Copernicus it did not

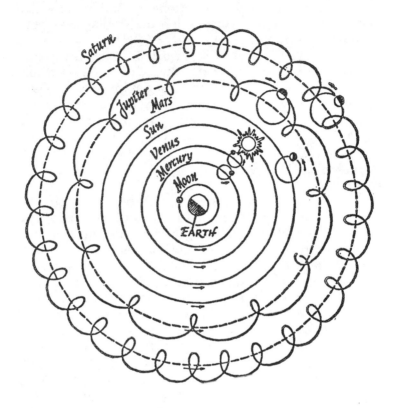

Ptolemy's Scheme of the planets' motions.

matter that satellites went around Jupiter; so long as
Jupiter was moving around the earth, the satellites par-
took of Jupiter's motion and did not disturb Ptolemy's
geocentric universe.)

It is easy to understand why Venus's phases fit the
Copernican system but are not compatible with the Ptolemaic. In this latter system Venus moves around a small
circle (epicycle) whose center moves in its turn, describing a circle around the earth. Venus would always turn
to the earth its non-illuminated parts, and an observer
on the earth would never see more than a thin crescent

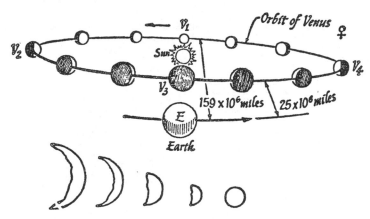

Venus in the Copernican System. The planet presents to the earth illuminated parts which change in shape. The size of Venus also seems to change as the planet moves around the sun. When going from V_3 to V_1, Venus is seen as the morning star, from V_1 to V_3 as the evening star. At the bottom are Galileo's own sketches of the phases of Venus as the evening star.

of it. But in the Copernican system Venus moves around the sun, *closer* to the sun than the earth is. Accordingly, when Venus is in V_1 and the earth in E (see the drawing above), Venus must appear, as Galileo actually found, "completely round, as morning star, but rather small in size." (In this position Venus would be hidden by the sun if the orbits of Venus and earth were on the same plane; they are, however, on planes forming an angle.) When Venus is in V_3 and the earth in E, Venus must turn its dark side to the earth and "will vanish." Because the period of revolution of Venus is about two thirds the period of the earth, Venus must go across the line uniting sun and earth faster than the earth and therefore invert its crescent shape in respect to the earth. Before passing over V_1 it will appear in the morning, shortly before sunrise; after going through V_1 it will shine in the sky at sunset time.

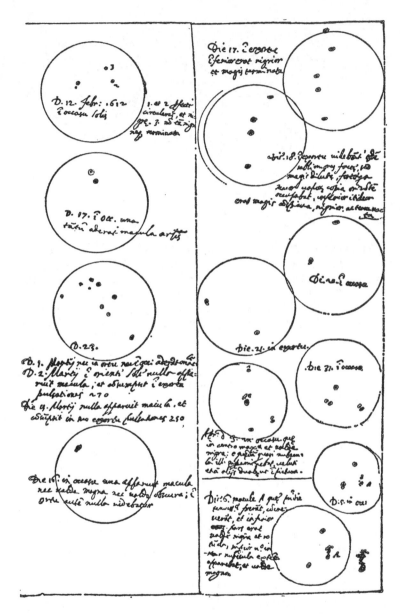

Galileo's Drawings of Sunspots. This is a reproduction of a page of his notes.

SUNSPOTS

All Galileo's astronomical discoveries mentioned so far were made in 1610. Shortly afterward he made another important observation, that of sunspots. We do not call it a discovery because sunspots had been observed before, probably even in ancient times, and a booklet mentioning them, by Johann Fabricius of Wittenberg, had appeared in 1611, before Galileo published their description. But Galileo seems not to have known of this booklet and said that he was the first discoverer of sunspots. So did another astronomer, the German Jesuit father Christopher Scheiner (1575-1650), and the two entered into a bitter dispute which played an important role in Galileo's controversy with the Church, of which we shall say more later. We are concerned here only with Galileo's observations. In his own words:

". . . the dark spots seen in the solar disk by means of the telescope are not at all distant from its surface, but are either contiguous to it or separated by an interval so small as to be quite imperceptible. . . . Some are always being produced and others dissolved. They vary in duration from one or two days to thirty or forty. For the most part they are of most irregular shape, and their shapes continually change, some quickly and violently, others more slowly and moderately. They also vary in darkness, appearing sometimes to condense and sometimes to spread out and rarefy. In addition to changing shape, some of them divide into three or four, and often several unite into one. . . . Besides all these disordered movements they have in common a general uniform motion across the face of the sun in parallel lines. From special characteristics of this motion *one may learn that the sun is absolutely spherical, that it rotates from west to east and around its own center, carries the spots along with it in parallel circles, and completes an entire revolution in about a lunar month.*" (Editors' italics.) Thus the sun itself, the most resplendent of celestial bodies, has

blemishes on its surface, and it rotates around its axis
as the earth and moon do.

THE ARISTOTELIAN UNIVERSE IS SHAKEN

News of the telescope and its marvels spread with a
rapidity unprecedented in the history of scientific discovery. Even 67 years after Copernicus died, he was
known only in restricted circles of mathematicians and
astronomers; Kepler's outstanding contribution to astronomy was not appreciated in his lifetime. But Galileo's
discoveries were the subject of discussion throughout
Europe within a few weeks of the time he made them.

Galileo himself greatly helped to spread the news by
talking and writing. It seems he could not keep his discoveries to himself for even a short while. He sent out
many scientific letters, and by March 1610 had finished
a book, *The Starry Messenger,* in which he described the
observations he had made so far. He wrote this book in
Latin because he wanted astronomers in foreign countries
to read it, and he certainly achieved his purpose. As
soon as Kepler read it, he had it reprinted in Germany;
five years after its publication it appeared in China, translated into Chinese—an astounding feat, if we consider the
slow means of transportation available at that time.

The great publicity that Galileo received was favorable
in most circles. Friends hailed him, poets wrote poems
in praise of his telescope, painters reproduced it with
their brushes, and innumerable people asked for a telescope or at least a chance to look through one. Among
scientists, however, there were many unbelievers. For one
thing, only Galileo could build telescopes powerful enough
to reveal all his discoveries, and he never gave away
the secret of how he ground his lenses. Though he built
a large number of telescopes, not all were equally successful; only ten were sufficiently good to show all the
new phenomena. Thus even Kepler doubted the existence
of the Medicean Planets at first, and only when he obtained a telescope made by Galileo did he admit that
Jupiter's satellites were real and not a misinterpretation

stemming from a flaw in the lenses or some other accident. He then wrote Galileo a letter ending: *"Galileae vicisti* [You have won, Galileo]."

More stubborn unbelievers were the Aristotelian philosophers, one of whom refused even to look through the telescope. Those who did look said either that they could see nothing new or that what they saw was mere optical illusion, tricks created by the telescope. Their mental bias is not hard to understand. They wanted to save at all costs the Aristotelian universe, which the telescope was shaking to its foundations.

In the universe of the telescope the sky contained many more stars than Aristotle had believed; his distinction between "terrestrial" and "celestial" bodies made no sense. The Medicean Planets circling around Jupiter formed a miniature Copernican system and disproved an argument of the Peripatetics: namely, that the earth could not be a planet, because it had a moon whereas the planets in the sky did not. Jupiter's satellites put the earth and Jupiter on the same level. The sunspots proved that perfection did not belong even to the sun, and, most important of all, the phases of Venus demonstrated that a planet *could* circle around the sun.

The tremendous blow that Galileo delivered to a well-rooted way of thinking, this opening of new vistas in a universe that had been thought unchangeable, is a greater contribution to astronomy than his discoveries. He could have gone far beyond and advanced the theory of astronomy had he not committed a serious mistake—he never really read Kepler's books.

GALILEO AND KEPLER

At about the time when Galileo built his first telescope, Kepler published a book, *Astronomia Nova* (*New Astronomy*), in which he presented the first two of his three laws. Kepler had been the assistant of the famous Danish astronomer Tycho Brahe (1546-1601), and upon Tycho's death he had acquired the numerous data on the positions of the planet Mars that his teacher had collected over 38 years.

It was in analyzing these data, which he knew to be very accurate, that Kepler found they did not fit a circular orbit. Further study led him to the conclusion that Mars and all other planets describe elliptical orbits, and he was able to determine how their speeds varied as they moved along their orbits.

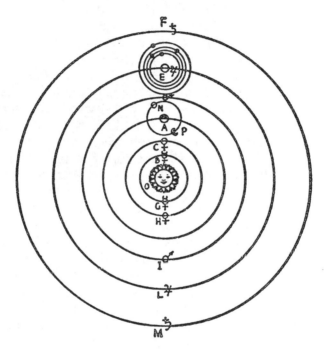

Galileo's Diagram of Copernicus's Solar System. It shows the Sun at the center and the planets moving on circular orbits around it. From the Sun outwards, the planets are: Mercury, Venus, the Earth and its moon, Mars, Jupiter and its four moons, and Saturn.

Galileo, like many other scientists in his time, did not notice Kepler's discoveries or pay attention to them. There was reason, if not justification, for this oversight. Kepler and Galileo were very dissimilar in temperament and aptitudes. While Galileo's intellectual powers spread over a wide range and he rightly could be called a complete man, Kepler was exclusively a theoretical astronomer, of very great mathematical ability. We should not be too far off if we said that Galileo was a man of great common sense and Kepler a man of genius. Kepler was profoundly mystical, and to his fanciful imagination not even astrology was alien. Mathematics and mysticism joined to give him a deep faith in a supreme harmony of the universe, whose laws should be expressed by simple and well-defined relations of proportionality. This faith is evident in the title of another book which he published a few years after *Astronomia Nova,* the *Harmonices Mundi,* or *Harmonies of the World.* He thought that the relations between the maximum and minimum speeds of each planet were harmonic in the musical sense and expressed by musical intervals. Thus, he said, planets played a music that only the angels could hear.

In Kepler's books discoveries were mixed with fancies of this kind and frequent metaphysical digressions, and, to make matters worse, they were written in prolix Latin. They were far from easy reading for anyone. They went against Galileo's rational and objective spirit. Despite their friendship, Galileo must have lost patience with Kepler's writings. "I have always held Kepler to be a free spirit (and perhaps too free) and keen," he once wrote. "But my way of philosophizing is most different from his; it has happened that writing about the same matters, only however when dealing with celestial motion, we have occasionally met with similar ideas . . . but this will not happen more than once in about one hundred of my thoughts."

Galileo never accepted the idea of planets describing elliptical orbits, although he probably became aware that Kepler had advanced this idea. In this Kepler was the more modern of the two, and Galileo, the founder of modern scientific thought, was trailing in the past, not able to free himself from the Platonic concept that perfect motion is circular. Galileo, who was trying all the time to understand how nature actually works, never realized that Kepler, seeking harmonic relations in the universe, had found the answer.

It is regrettable that Kepler and Galileo never met, never had a chance to discuss their work together. A chain of mountains, the Alps, and the barriers of language kept them apart. About 35 years after Kepler's death, Isaac Newton (1642-1727) abolished the barriers and at long last put their works together. But it took the passage of time and the great genius of Newton to clarify the dynamical concept of force. Once this concept was clear, Newton could achieve the synthesis of Kepler's astronomy and Galileo's work on motion. The result was his universal law of gravitation.

6. Florence and Rome

WHEN, AFTER THE INVENTION OF the telescope, the Venetian Republic confirmed Galileo's professorship for life, he might well have been fully satisfied and content. Not so. In the letter, already mentioned, that he wrote to his brother-in-law Landucci, the note of nostalgia is plain: "I find myself bound for life, and I shall have to be content with enjoying my *patria* a few times during the summer vacation."

Galileo's use of the word *patria* ("fatherland") is revealing. Florence and Padua are less than 150 miles apart; people in the two cities speak and always spoke the same language, with only some dialectical variations. Yet in Padua, Galileo felt an exile and was homesick. Despite recognition and honors, despite the stimulation of a brilliant life perfectly tuned to his spirit, he longed for Florence. He longed for the quiet that pervades the cities of Tuscany and the soft hills around them, that seems to have pervaded them from the beginning of time.

He was about 45 years old and had reached the time of

life when one wants to look back and consider one's achievement, to complete the unfinished and perfect the finished. He wanted to make further astronomical observations, especially of the Medicean Planets, and to conclude his studies of falling bodies and of the motion of projectiles, which he had never given up since the days of Pisa. He did not want to keep his knowledge to himself but wished to share it with "the entire world," and to this end he planned to write many books. It would take time. In Padua, he felt, he did not have much spare time, because fame brought pupils and the curious flocking to his door; lectures and private lessons filled most of his days.

For some time he had been playing with the idea of returning to Florence to be under the protection of the Grand Duke, his former pupil Cosimo, who had treated him with reverence and affection. Even before the invention of the telescope, Galileo had corresponded with the court of Tuscany about a position there. "It is impossible to obtain wages from a republic, however splendid and generous, without having duties attached," he had written in February 1609. "For to have anything from the public one must satisfy the public and not one individual. And so long as I am capable of lecturing and serving, no one in the Republic can exempt me from duty while I receive pay. In brief, I can hope to enjoy these benefits only from an absolute ruler."

It may seem strange that a man who strenuously advocated freedom of thought in science could accept, even seek, absolute authority in government. Such contradictions are common in men who live in the great periods of transition. In politics Galileo, the first modern physicist, was far from the liberalism which, strengthened by the scientific progress and rational thinking to which he greatly contributed, was to develop a century later and eventually bring such sweeping changes as the American and French revolutions.

The struggle to free himself and scientific thinking from medieval ties absorbed Galileo completely. In politics he accepted contemporary trends. His allegiance to his ex-pupil Cosimo II was not only the affection of a teacher but also the deference of a subject. The Florence over which Cosimo ruled was the proper home, the *patria,* for as faithful a subject of the Medici as Galileo.

His telescope, his dedication of *The Starry Messenger* to Cosimo, and his glorification of the Medici in the naming of Jupiter's satellites facilitated his return to Florence. Soon he obtained the position he sought. He was appointed First Mathematician at the University of Pisa, with no obligation to lecture or reside there, and personal Philosopher and Mathematician to the Grand Duke Cosimo. In September 1610, only a year after he had gratefully accepted his lifetime appointment in Padua, he went back to Florence. Perhaps he realized that this offended the noble Venetian rulers who had welcomed him when he was an obscure and needy teacher and had later given him prizes and honors for his scientific achievements; in any event, he never went back to either Padua or Venice, even for a brief visit.

There rarely is much to be gained from speculating what might have happened if. . . . Yet in the case of Galileo's decision to return to Florence one must wonder. It seems plausible to assume that if he had stayed in Padua, the rest of his life would have been much more calm and secure.

Venice was then one of the European states recognizing the freedom and human dignity of all its citizens. It did not give anyone "wages without duties attached." Venice did not tolerate interference in its internal affairs, even from the Church. Indeed, only four years before Galileo quit Padua the Republic had defied Pope Paul V in a struggle over the Republic's right to try priests in civil courts. When the Pope placed Venice under interdict and threatened war, the Republic retaliated by expelling the Jesuits and two other religious orders.

In Florence the situation was different. There the policies of the Medici, who reckoned cardinals and popes in their family, were tightly knit with those of Rome. In the previous century the Church had been seriously threatened by the Reformation and the spread of Protestantism. It reacted to this threat with a long and intense campaign against Protestantism and all other forms of what it considered heresy. This was part of a prolonged and sustained effort of the Roman Church to strengthen itself by many measures, including removal of abuse from within and clearer definition of the doctrines called in question by the Reformers. This reaction by the Roman Church is known as the Counter-Reformation. It lasted over a century, and roughly speaking, it coincided with Galileo's lifetime.

SAGREDO'S PROPHECY

The conservative currents of the Counter-Reformation affected not only religion but also politics, especially in countries closely tied to the Church, like the Grand Duchy of Tuscany. This state of affairs did not bode well for a man of Galileo's background and forthright temperament, and it worried at least one of his closest friends, the young Venetian nobleman Sagredo, of whom we already have spoken.

Sagredo, who for his culture and wisdom could be considered a true exponent of the Venetian ruling class, was thoroughly familiar with the social and political structures of the Europe of his day and sensed the implications for Galileo. He wrote Galileo a letter that, in view of what was to happen, sounds almost prophetic: "Where will you find freedom and sovereignty of yourself as in Venice? Especially since you had here support and protection, which became more weighty every day as the age and authority of your friends grew. . . . At present you serve your natural Prince, a great, virtuous man of singular promise; but here you had command over those who command and govern others, and you had to serve only yourself. You were as the ruler of the universe. . . . In the tempestuous sea of a court who can be sure not to be . . . belabored and upset by the furious winds of envy? . . . Who knows what the infinite and incomprehensible events of the world may cause if aided by the impostures of evil and envious men . . . who may even turn the justice and goodness of a prince into the ruin of an honest man? I am very much worried by your being in a place where the authority of the friends of the Jesuits counts heavily."

Sagredo, who was Venetian consul in the Levant when Galileo left Padua, wrote the letter upon his return. By the time it reached its destination, the chain of events that was to lead Galileo into the controversy with the Church had already started.

Galileo was extremely religious not only by education but also because in the life of all creatures—in all natural phenomena—he saw signs of a Supreme Will and Power, inscrutable but perfect. In his lucid mind he may have experienced some conflict between his reasoning as a scientist and his loyalty to the Church. But this loyalty was very strong, and ironically it was one of the causes that led him into the controversy with the Church. He knew that a revolution in man's conception of the physical universe was at hand and that dogma alone could not stave it off. As a good Catholic, he did not want his church to take a doctrinal stand from which it would have to retreat. He was an optimist and believed that he could persuade even the most reluctant to accept his proofs. His persistence brought about a drama that went way beyond its origins and exceeded the stature and the personal value of its actors.

By the time he returned to Florence, Galileo was sure that he had the proofs of the Copernican system, and to him the idea that the sun stands still while the earth and planets revolve around it no longer was a likely hypothesis but a definite fact. He knew that the theologians would not accept this readily, but he also knew how powerful and convincing his arguments could be, how strong the evidence of the telescope was to anyone of good faith. So in his optimism he resolved to go to Rome and convert the theologians.

TRIUMPH IN ROME

Galileo's visit to Rome in the spring of 1611 was a triumph. Wherever he went he was feted. He met high prelates and cardinals, among others Robert Cardinal Bellarmine, Papal Secretary of State, then the foremost Catholic theologian and one of the Inquisitors General, whom the Church has made a saint. Pope Paul V received Galileo with benevolence, a sign of great favor, since the quiet, reserved Pope did not readily grant audiences. Galileo made friends among noblemen, artists, and writers. On the fair nights of that Roman spring he set up his telescope in lovely gardens of old palaces, showed his discoveries to large numbers of enthusiastic people, and "converted unbelievers one after the other."

He also received a signal honor, of which he was to be exceedingly proud the rest of his life: He was elected a member of the Accademia dei Lincei (named for the lynx, which has piercing eyes). It had been founded a few years before by the young Roman Prince Federico Cesi. Among the many academies then sprouting over Europe as a reaction against the pedantry and conservatism of universities, this was the first that specialized in science and philosophy, and Galileo was its sixth member. (The Lincean Academy rapidly expanded and, like many similar institutions that came after it, was closely associated with the development of science.)

The extent of Galileo's Roman triumph is set forth in a letter a cardinal wrote to the Grand Duke. Galileo, he said, "had occasion to demonstrate his inventions so well that they were held by all capable and expert men to be not only very true and real, but also most wonderful, and were we now in that ancient Roman republic, I trust that a statue would be erected to him on the Capitol to honor his excellence and valor."

This lively response, the interest he had aroused, certainly flattered and pleased Galileo, who always enjoyed recognition. Everybody talked enthusiastically of his astronomical discoveries. He had converted "unbelievers." He brought discussion of the Copernican system and its implications out of the restricted fraternity of mathematicians and astronomers into much larger circles of cultivated people—even, indeed, into the papal court.

But Galileo's success could also arouse alarm in some members of the papal court, where novelties, and especially those touching the heavens, always caused concern. Cardinal Bellarmine had looked through Galileo's telescope, had seen the novelties with his own eyes, and wondered what might come of them. Cardinal Bellarmine was a *consultor* of the Holy Office, or Inquisition. When Galileo, riding the high tide of his triumph, returned to Tuscany, he did not know it, but the Inquisition was alerted.

THE INQUISITION

We have mentioned the Inquisition several times, and we must now say a few words about it, which will help the understanding of subsequent events. The Inquisition was the high tribunal of the Roman Church. It had functioned through the Middle Ages, but in its current form it dated from some 20 years before Galileo's birth. During the Counter-Reformation the Inquisition played a very important role. Working always in secrecy, through its representatives in Catholic cities, it spotted, investigated, and prosecuted those suspected of heresy, trying to suppress it before it could spread. It also exerted censorship to avoid the perversion of Catholic doctrine, and to this end it collaborated with the Congregation of the Index, which published the *Index,* or list of prohibited books. Inquisitors, and other "licensers" as well, granted the permission, or *imprimatur* (literally, "be it printed"), without which no book could be published.

To have a book placed on the *Index* was not too serious a mishap. It happened to many good Catholics, and Cardinal Bellarmine himself barely escaped it: Pope Sixtus V ordered the Cardinal's book, *De Controversiis,* to be placed on the list of prohibited books, but the Pope died shortly afterward, and Bellarmine's book did not stay on the *Index*.

To be tried by the Inquisition was something that nobody could take lightly. Although in Galileo's time the Inquisition was becoming more and more lenient, it was known to have used torture in the past and to have sent many heretics to burn at the stake. As late as 1600 this fate had befallen the Italian thinker Giordano Bruno, a one-time Dominican friar who had adopted a pantheistic philosophy of nature.

POLEMIC WRITINGS

Soon after Galileo's return to Florence from Rome in the early summer of 1611 the Grand Duke invited him to a dinner party at his court. This was not unusual, as Cosimo was bound by affection to his mathematician. Besides, he liked to gather around him the best scholars in Tuscany, especially when he had illustrious guests, and hear them talk about the issues of interest to the learned world. At this dinner the discussion centered on floating bodies. Galileo maintained that bodies can float only if their specific gravity is less than that of water. Among the dinner guests there were, however, some Peripatetic philosophers, and they argued that bodies float if their shape is wide and smooth so they cannot cut through the resistance of the water. Floating bodies were a subject in which Galileo was especially well grounded; it had interested him since he had read Archimedes in student days and written *The Little Balance*. He was able to uphold his point so aptly and brilliantly that one of the guests of honor took sides with him. This supporter was no less than Maffeo Cardinal Barberini, who years later was to become Pope Urban VIII. He was also going to turn against Galileo and become one of his b'tter enemies, but for the moment he was as affable as one could be, sincerely admiring Galileo's dialectical skill. Perhaps to please this cardinal, the Grand Duke asked his mathematician to put his argument into writing. A short polemic book was the result: the *Discourse on Floating Bodies*.

Galileo's sharp, almost sarcastic wit made him especially suited to arguments and debates, of which he was to have many in the following years. Some of these resulted in famous polemic writings that added to his lasting glory; all were to antagonize people, turn them into those "enemies" whom he later blamed for all his difficulties and misfortunes.

The Peripatetics at the Grand Duke's table were not very dangerous as potential enemies, but his next adversary was different. Even before the *Discourse on Floating Bodies* was published in 1612, Galileo was engaged in a dispute with an astronomer whose name he did not know and was not to find out for over a year, the Jesuit father Christopher Scheiner, whom we have already mentioned. This controversy arose from Galileo's ideas about the sunspots which he had observed in 1610 and the discovery of which both he and Father Scheiner claimed. Father Scheiner had also made observations of sunspots and communicated his opinions in several letters to Mark Welser, a German patron of science and a Lincean academician. Perhaps to avoid direct criticism, Scheiner wrote under a pen name. Mark Welser published Scheiner's letters and sent them to Galileo for comment without revealing the name of the author.

Galileo replied in three *Letters on Sunspots* addressed to Welser (in Italian, which Scheiner could not read and had to have translated, while Scheiner had not written in his native German but in Latin). In his letters Galileo criticized Scheiner's views: that spots were caused by small planets removed from the body of the sun and revolving around it; that the spots moved from east to west, slanting from north to south (while Galileo had observed their motion to be in the opposite direction); that they could not be on the surface of the sun because it was unbelievable that dark spots existed on a "most pure and most lucid body"; that they were blacker than spots on the moon. His criticism of these and other erroneous opinions gave Galileo occasion to review most of his astronomical observations, especially, and in greater detail, those on sunspots. The paragraphs on sunspots that we quoted when describing what Galileo had seen with the telescope come from these *Letters*.

NATURE AND THE SCRIPTURES

For the complicated story of Galileo's relations with the Church the greatest significance of these *Letters on Sunspots* lies in the fact that for the first time in print he openly endorsed the Copernican system as a reality and not a mere hypothesis, and that he used his own discoveries as proofs in favor of Copernicanism. Important also is the fact that unwittingly he antagonized a Jesuit, the first of many against whom he had occasion to write over the years. The Jesuits were powerful in the Church, and in particular they were advisers on educational matters. It was unfortunate indeed that so many of them sooner or later should withdraw their previous friendship, respect, or perhaps indifference toward Galileo to pass into the "enemy's" camp.

Trouble, however, came at first from other quarters. Here is what happened. In 1613 Galileo learned from Father Benedetto Castelli, one of his most beloved pupils, that in the course of a discussion at the court of Tuscany the Dowager Grand Duchess, Cristina di Lorena, had taken the stand that the earth could not move, because its motion would contradict the Holy Scriptures.

Galileo decided that the time had come to explain his views on the relations between science and faith. He did this in his *Letter to Castelli,* which he sent, in manuscript copies, not only to his pupil Castelli but also to several friends. Soon afterward, in his *Letter to the Grand Duchess Cristina,* he elaborated what he had written to Castelli, and the letter has become rightly famous. It is lofty and solemn and shows that Galileo's faith in nature and its laws went side by side with his faith in God. It contains passages which are among Galileo's most beautiful. "I think in the first place," Galileo wrote, "that it is very pious to say and prudent to affirm that the holy Bible can never speak untruth—whenever its true meaning is understood. But I believe nobody will deny that it is often very abstruse, and may say things which are quite different from what its bare words signify."

And further on: "Nature . . . is inexorable and immutable; she never transgresses the laws imposed upon her, or cares a whit whether her abstruse reasons and methods of operation are understandable to men. For that reason it appears that nothing physical which sense-experience sets before our eyes, or which necessary demonstrations prove to us, ought to be called in question (much less condemned) upon the testimony of biblical passages which may have some different meaning beneath their words. For the Bible is not chained in every expression to conditions as strict as those which govern all physical effects; nor is God any less excellently revealed in Nature's actions than in the sacred statements of the Bible. . . .

"But I do not feel obliged to believe that that same God who has endowed us with senses, reason, and intellect has intended to forgo their use and by some other means to give us knowledge which we can attain by them. He would not require us to deny sense and reason in physical matters which are set before our eyes and minds by direct experience or necessary demonstrations."

In our day these views are widely shared and officially recognized by the Church. More than two and a half centuries after Galileo, in 1893, Pope Leo XIII wrote an encyclical entitled *Providentissimus Deus* which gave the Church's official point of view concerning the relations between science and Scripture. This statement cannot be distinguished from Galileo's.

Even in Galileo's time, his letters to Castelli and Cristina were never called in question by the highest authorities of the Church, as we shall see. But some in the Church did criticize them. To some of those who accepted the Bible as a privileged source of knowledge and had little understanding of the new developments in science, Galileo's writings seemed an outsider's interference in religious matters. A Dominican friar denounced the *Letter to Castelli* to the Inquisition. Another Dominican, Father Tommaso Caccini, who had once been disciplined for being

a "scandal-maker," preached a sermon against Galileo in
the popular church of Santa Maria Novella in Florence.
He ended by saying that mathematics was an art of the
devil, that mathematicians were the source of all heresies
and should be ousted from all countries. Shortly afterward he, too, testified against Galileo before the Inquisition.

Through Cardinal Bellarmine the Inquisition, we have
seen, was aware of Galileo's campaign in favor of the
Copernican doctrine as early as 1611, yet it is certain
that at that time the matter was not serious. But now the
Inquisition had formal charges against Galileo. Despite the
secrecy surrounding the Inquisition, Galileo must have
been sufficiently aware of what was going on to decide
that his presence in Rome was needed.

Upon his arrival late in 1615, he received the same
warm welcome as the time before. (As on the previous
visit, he stayed at the Villa Medici, the Tuscan embassy,
high on the Pincio Hill overlooking the domes of the
many churches; and on the Grand Duke's order he had
"board for himself, a secretary and a small mule.") He
at once resumed his campaign in favor of Copernicus,
intensely, through talks and discussion with almost ev-
eryone of importance in Rome and through several new
writings. He took a much more definite stand and stated
that Copernicus himself had never considered his system
as a mere scientific hypothesis but as reality, as unmis-
takable fact. Could people not see this? It was most evident in Copernicus's own writings, if only people understood them.

Galileo had enemies busily denouncing him to the Inquisition, but he also had friends at court. Several cardinals did their best to persuade him to keep quiet in public
about Copernicus, regardless of his private belief. Had
he heeded their advice, he probably would have avoided
his subsequent troubles and perhaps even changed the
course of the Church's history. For until he popularized
the Copernican theory, the Church, as represented by its
chief theologian, had by no means closed its mind.
Cardinal Bellarmine wrote a letter to a friend of Galileo,

the Carmelite monk Paolo Antonio Foscarini, in which he said, "I say that if a real proof be found that the sun is fixed and does not revolve around the earth, but the earth around the sun, then it will be necessary, very carefully, to proceed to the explanation of the passages of Scripture which appear to be contrary, and we should rather say that we have misunderstood these than pronounce that to be false which is demonstrated."

THE DECREE AGAINST
THE COPERNICAN DOCTRINE

But Galileo kept talking in favor of the Copernican doctrine, and while he busied himself at this, the Inquisition went on examining his case. It never really questioned the theological views that Galileo had expressed in his letters, and although an early formal denunciation charged him with heresy and blasphemy concerning the nature of God, he was able to clear himself of those accusations. But on February 23 a group of eleven theologians was convened to examine his beliefs in science, or, rather, what the Inquisition *said* were his beliefs. Two propositions were submitted to the theologians: (1) "The Sun is the center of the world and hence immovable of local motion." (2) "The Earth is not the center of the world, nor immovable, but moves according to the whole of itself and also with a diurnal motion." The language here is not scientific and not Galileo's, and the two propositions were not quoted from his writings. The following day the theologians decreed that the first proposition was "foolish and absurd, philosophically and formally heretical"; that the second could be "equally censured philosophically and was at least erroneous in faith."

This decree was followed by two important events, neither of which the trusting Galileo had expected in the least. On the Pope's instruction, Cardinal Bellarmine called Galileo to his palace. There, in the presence of witnesses, he admonished Galileo that he was not to hold, teach, or defend the condemned opinion of Copernicus.

A few days later the second event took place: years after its publication, the book *De Revolutionibus,* which Canon Copernicus had dedicated to a Pope, which the Pope had accepted, and with which the Church had found no fault until Galileo had started to present it as a reality, was condemned and prohibited until it should be corrected. Not that Copernicus had never been criticized. Martin Luther had said: "People give ear to an upstart astrologer who strove to show that the earth revolves. . . . This fool wishes to reverse the entire science of astronomy; but sacred Scripture tells us that Joshua commanded the sun to stand still, and not the earth." And John Calvin had written in the same mood: "Who will venture to place the authority of Copernicus above that of the Holy Spirit?" Yet, the Roman Catholic Church had taken no action against Copernicus's books or his ideas until Galileo undertook his campaign to "convert" the theologians.

Galileo was too intelligent not to have at least some realization of the part he himself had played, of how his astronomical observations and the very force and persistence of his arguments had provoked this condemnation. Copernicus's book had been presented as a mathematical hypothesis. It had been understood by few and had been considered in the same class with the many intellectual exercises in which for centuries philosophers and poets had indulged in the hope of adding something to the Aristotelian system. But at the hands of Galileo the heliocentric system was threatening the geocentric and, much more serious, God's creation was becoming an object of direct human observation, which could be interpreted without the help of the Scripture or of religion.

The blow fell heavily on Galileo. For some time he had planned to write a book on the two systems of the world, Ptolemaic and Copernican, an "immense design full of philosophy, astronomy and geometry," and his friends had been impatiently waiting for it. Now he could not write it. Now he was silenced.

Indeed, for a long time his powerful voice did seem stilled. For ten years after the publication of the *Letters on Sunspots* nothing appeared in print over his name. This is not to say that he remained idle. He negotiated with Spain over navigational uses of his astronomical discoveries, and he worked on a telescope primarily designed for sailors. He pursued his astronomical observations and the other studies that he had never really given up. All this, for Galileo, could mean only marking time.

Then in the summer of 1618 three comets appeared in the sky. This new astronomical event was too great a challenge. Galileo, the silenced, could not resist the temptation to make himself heard again. He was seriously ill and unable to observe the comets, but his friends kept him informed. The comets aroused great interest in the scientific community, and a Jesuit astronomer, Father Horatio Grassi, cited them as the best proofs against the Copernican doctrine. This was too much. Galileo's friends urged him to reply. With Cardinal Bellarmine's admonition in mind, Galileo preferred not to discuss astronomy in public. Instead, he had his pupil Mario Guiducci lecture on comets and publish a *Discourse on Comets,* which actually set forth Galileo's own views.

THE ASSAYER

Father Grassi, writing under the name of Lothario Sarsi, replied with a paper called *The Astronomical and Philosophical Balance*—a "balance" with which he professed to "weigh" Galileo's opinions as presented by Guiducci. He made a criticism of Galileo's views which Galileo thought unjust. The scientist therefore decided eventually "to break his previous resolve to publish no more" and came out with *The Assayer,* a tract which is now considered the best polemic ever written in the Italian language. Galileo had sharpened his pen and wits and attained proficiency in this sort of writing in his previous disputes, not only those we have mentioned but also less famous ones.

The full title of the booklet has a quaint flavor: "The Assayer, where with an Exquisite and Fair Balance the

Things are Weighed which are Contained in the Astronomical and Philosophical Balance of Lothario Sarsi of Signa, written in Form of a Letter to Monsignor Cesarini ... [Cesarini's titles follow] ... by Signor Galileo Galilei, Lincean Academician, Florentine Nobleman, Philosopher and First Mathematician of the Most Serene Grand Duke of Tuscany."

The Assayer's balance is not as fair as Galileo wanted people to believe, and besides, in the light of modern knowledge, not many of his views on comets appear correct. To us this is not the main point. In *The Assayer,* as in his other polemics, Galileo asserted his right to use his own brain and showed how masterfully he could use it. He placed most problems in their proper perspective, analyzing them rigorously and drawing the inevitable logical conclusions from them.

The Assayer, like his other polemic writings, is full of statements against "the firm belief that in philosophizing one must support oneself upon the opinion of some celebrated author," and against those who "judge a man's philosophical opinions by the number of his followers." "The crowd of fools who know nothing, Sarsi," Galileo wrote, "is infinite. Those who know very little of philosophy are numerous. Few indeed are they who really know some part of it, and only One knows all."

Thus we see that the dispute between Father Grassi and other Jesuits on one side and Galileo on the other was not really about comets but about a way of thinking. It was again the gathering conflict between authority and the intellectual freedom of science.

The following excerpt from *The Assayer* is both enlightening and amusing:

"If Sarsi wants me to believe ... that the Babylonians cooked their eggs by whirling them in slings, I shall do so; but I shall say that the cause of this effect is very far from what he suggests [friction of the air]. To find the true cause I shall reason thus: 'If we do not achieve an effect that others formerly achieved, then it must be that in our doing we lack something that was the cause of their success. And if there is just one single thing that we lack, then that alone can be the true cause. Now we do not lack eggs, nor slings, nor sturdy men to whirl them; yet our eggs do not cook; on the contrary, they cool down faster if they are hot. And since nothing is lacking to us except being Babylonians, then being Babylonians is the cause of the hardening of eggs, and not friction of the air. This is what I wished to prove.' Is it possible that Sarsi has never noticed what coolness the continuous change of air brings to his face when he is riding post-haste? And if he has, will he prefer to believe things happened over two thousand years ago in Babylon and related by others rather than present things which he himself experiences?"

Between the lines of Galileo's witty passage we read the same faith in nature and its laws that he expressed repeatedly, occasionally in joking tones, more often in solemn, almost religious terms.

7. Galileo and Urban VIII

AN UNEXPECTED EVENT that took place in 1623 while the Lincean Academy was printing *The Assayer* seemed to clear the air for Galileo. Maffeo Cardinal Barberini, whom Galileo regarded as his friend and who had sided with him in his dispute about floating bodies, was elected Pope under the name of Urban VIII. Maffeo Barberini belonged to a noble Florentine family, friends of the Grand Dukes. He was a patron of arts and letters who had written and published poems, a Lincean academician, and Galileo's sincere admirer. The cardinal liked to consider himself a philosopher and had often discussed science with Galileo, both in Florence and in Rome. Indeed, as a token of his admiration he had written a Latin poem in Galileo's honor. After becoming Pope, he eagerly asked Prince Cesi, who had gone to pay him homage, whether Galileo also was coming. The auspices could not have been more favorable.

Urban VIII, the Pope, however, was to reveal less attractive traits that had not been evident in the cardinal. Shrewd, with an extremely high opinion of himself, avid for power, jealous of his authority, he was so ambitious both for himself and for his family that his nepotism was to become proverbial. The Barberini's revenues and their power became enormous. They adorned their homes and palaces sumptuously, and popular opinion accused them of having stripped Rome's monuments of the few marbles and ornaments the barbarians had not removed long before.

Hence the saying in Italy, where sayings may still be in Latin: *"Quod non fecerunt Barbari fecerunt Barberini* [What the barbarians did not do, the Barberini did]."

Galileo was very anxious to go to Rome and kiss the slipper of his old friend, the new Pope. He hoped again that the Copernican doctrine might be accepted and that he might obtain permission to write about it. Illness delayed him. (Since his return to Florence from Padua, illness had bothered him repeatedly.) At last he went to Rome in the spring of 1624. Urban VIII received him in many friendly interviews, highly praised *The Assayer,* which Galileo had dedicated to Urban, gave him presents and promise of a pension. So far as the Copernican theory was concerned, Cardinal Barberini could well listen to quaint ideas and even try to understand them, but Urban VIII, the Pope, spoke for the Roman Church, for his flock of the faithful throughout the entire world. He agreed—or did Galileo, the ever-trusting, see things rosier than they were?—that Galileo could write about Copernicus's system, provided he represented it not as reality but as a scientific hypothesis.

THE DIALOGUE

By June, Galileo was back in Florence, anxious to start work on the book for which he had "philosophized" all his life, the "immense design" which he had announced to his friends when leaving Padua. He was 60 now and in poor health. How many years did he still have to write? Yet he worked slowly, as if in doubt. He put the book aside to resume experiments; then illness visited him again, an illness so serious that he almost died. *Dialogue of the Two Greatest Systems of the World* was completed at last in 1630. In Rome, Galileo, not without difficulty and with much help from the Tuscan ambassador, obtained the *imprimatur* when he promised to correct the introduction and the ending paragraphs to make it clearer that the Copernican theory was only one of many possibilities.

The *Dialogue*, published in Florence in 1632, is in Italian, like most of Galileo's works. He had once explained why he did not write in Latin: "I am induced to do this by seeing how young men are sent through universities at random to be made physicians, philosophers and so on. . . . Other men who would be fitted for these [professions] are taken up by family cares and other occupations remote from learning. . . . Because they cannot read Latin . . . they believe that in 'those awful books' . . . there are things way above their heads. Now I want them to see that just as nature has given to them, as well as to philosophers, eyes with which to see her works, so she has given them brains capable of understanding and penetrating them [those works]."

The title page of the *Dialogue* states that "in the meetings of four days conversation is held about the Two Greatest World Systems, Ptolemaic and Copernican, and the philosophical and natural reasons for both are advanced *inconclusively*." *

The conversation is among three gentlemen, two of them representing friends of Galileo who had died a few years before—the Venetian Giovanni Francesco Sagredo, whom we have mentioned, and the Florentine Filippo Salviati (1582-1614). Salviati was "a sublime intellect who enjoyed no other pleasure more avidly than deep specu-

---

* Italics ours. The way in which Copernicus's ideas are presented is hardly inconclusive.

lation," the closest intellect to his own that Galileo had ever found. Galileo had often been a guest in Salviati's beautiful Villa delle Selve near Florence, where with Salviati's assistance he had performed many astronomical experiments. Salviati's untimely death in 1614, like Sagredo's in 1620, had deeply grieved him.

In the *Dialogue,* Salviati represents the scientist, and through his mouth speaks Galileo himself. (But when Galileo wanted to *stress* his own point of view he made his characters refer to "our friend the Academician.") Sagredo represents the intelligent and cultivated layman interested in science, and Galileo drew him from life. (A mutual friend wrote to him, "How skillfully you gave life to that worthy man, Sagredo. So help me God, I thought I heard him talk again!") The third character represents the Aristotelian philosopher who stands for authority and does not admit arguments that he cannot deduce from texts. Appropriately, Galileo gave him the name of an ancient commentator on Aristotle, Simplicio, a name, however, that contains the implication of "simple-minded."

Through the mouths of his three characters Galileo discusses almost everything he has studied and learned, not only his astronomical discoveries but also, for the first time in a published work, the results of 40 years of his study of motion. The form of dialogue gives him the best opportunity for showing his skill in comparing opposite views. The intellectual fight is between Salviati and Simplicio, while Sagredo's role is to admire constantly the wonders of nature and to bring additional ammunition against Simplicio's arguments. If Simplicio is not always willing to admit defeat, the reader, on the other hand, is left with no doubt: Salviati is always right, and the only true doctrine is the Copernican.

In all fairness, to any praise of the *Dialogue* we must add a criticism. Galileo believed that the most stringent proof of the Copernican system was his theory of tides, which he explained through the earth's two motions, around its axis and around the sun, and through the size of sea basins. Galileo considered basins like that of the

Mediterranean Sea as "vessels containing the water." When any vessel containing a liquid moves with non-uniform velocity, sometimes faster, sometimes more slowly, he explained, the water inside "may acquire a faculty of fluctuating," as in tides. He believed that the combined effects of the earth's rotation around its axis and revolution around the sun did in fact give sea basins this varying motion.

This theory is, of course, entirely erroneous. To Galileo it was a proof of the Copernican system because it implied that the only possible cause of the tides was the combination of the motions attributed to the earth in the Copernican system.

THE SUMMONS TO ROME

A few months after the publication of the *Dialogue* the publisher received an order from Rome to suspend its sales, and, much more ominous, a few months later Galileo received a summons to appear before the tribunal of the Inquisition in Rome.

He was dumbfounded. He had complied with all instructions, made all the changes the authorities had requested. The book had been carefully examined in Rome first, then in Florence, and had been licensed in both cities. No one was able to suggest what had gone wrong, and even today, three centuries later, and after a study of the once secret documents of the Inquisition, the story is puzzling.

One fact stands out: Pope Urban VIII was angry, and from a friend had turned into Galileo's bitterest enemy. Presumably he believed that Galileo had fooled him. The closing paragraph of the *Dialogue* contained a statement that he himself had suggested: The Copernican doctrine is "neither true nor conclusive" and "it would be excessive boldness for anyone to limit and restrict the Divine Power and Wisdom to one particular fancy of his own." But Galileo had put this statement in the mouth of Simplicio, the "simple-minded," the man who intellectually cuts the poorest figure throughout the *Dialogue.*

Galileo had had no choice in this, because it was Simplicio who upheld the anti-Copernican position. But it seems that someone hinted that in Simplicio Galileo had represented the Pope himself.

Actually it is highly improbable that Galileo made fun of the Pope intentionally, for he had always shown great respect for authority, had often sought the favor of the powerful, and at times had gone to considerable pains to obtain it. On the other hand, it seems just as improbable that in writing and revising his *Dialogue* he failed to realize that he was complying only with the form, rather than with the substance, of the licensers' requests—that he was actually presenting the Copernican doctrine as a proved fact rather than as an hypothesis.

However that may be, the licensers, not being astronomers, had not detected the strong Copernican slant in Galileo's book. Undoubtedly persons who saw this slant now took pains to point it out to the Pope. After suspicion became rooted in his mind, the Pope took the stand that the *Dialogue* was potentially more dangerous to Christianity than Luther's or Calvin's heresy. This view was suggested to him, it was widely thought at the time, by certain Jesuit quarters. Galileo had offended Fathers Scheiner and Grassi by his polemics and many other Jesuits by passages in the *Dialogue;* perhaps they were taking their revenge. Thus we see many possible reasons for Galileo's summons before the Inquisition: on the one hand, the wounded pride, personal ambitions, and vanities of his enemies, and on the other hand, the ambition and naïveté of Galileo himself, who had not been content with having his ideas accepted among scientists but had sought approval and recognition from the Church, hoping with the force of his words to overthrow traditions of thousands of years.

But there was more than a clash of personalities. Behind this clash there was a conflict between the long-standing traditions and authority of the Church and the need for freedom of thought in science. Had the men and the circumstances been different, the conflict might have been avoided in Galileo's time, but sooner or later a dis-

sension was inevitable. Objective science, with its ir-
resistible force of persuasion, would gradually have conquered modern society. Scientists would have pressed
for a prompt acceptance of their discoveries, but the
Church would not have easily accepted views that went
against its authority and its many-centuries-old traditions. Theologians trained in these traditions would have
felt it their duty to be cautious and to question the scientific proofs that seemed evident to scientists.

As it happened, the conflict between science and theology gathered so early and swiftly that the Church took
a strong action against the very founder of modern science. The way in which Galileo reacted to this action is a
proof of his own inner conflict. Had he been less devout
a Catholic, he could have refused to go before the Inquisition in Rome; Venice offered him asylum, and he
could have accepted it. On the other hand, had he been
less convinced of his scientific opinions, and of his duty
to defend those opinions, he could have given them up
and remained at peace with the men representing the
Church in his time. But Galileo could not resign himself
to take either course.

THE TRIAL

After receiving the summons to Rome, Galileo pleaded
for time and clemency, in view of his great age and poor
health and of a plague that had broken out in Tuscany,
making travel hazardous. But the Pope was inflexible,
and Galileo received a new injunction. Then he fell seriously ill. The Pope said that if his life was in danger, as
doctors stated, he could delay his trip, but as soon as he
recovered he would be brought to Rome in chains. The
Grand Duke Cosimo had long been dead, and his suc-
cessor, Ferdinand II, did not take as strong a stand
in defense of his mathematician as Cosimo might have.
He advised Galileo to comply and depart for Rome. Sagredo's prophecy was coming true.

Galileo arrived in Rome early in February 1633 and
stayed there five months—five long, trying months filled

with alternate hope and distress, with questionings by the Inquisitors and periods of suspense. The purpose of the trial was to ascertain whether Galileo was still holding the prohibited doctrine and whether in writing his *Dialogue* he had disobeyed orders. The main legal question was: Had Galileo in 1616 received only Cardinal Bellarmine's *admonition* "not to hold, defend, and teach the Copernican doctrine"? Or had he also received a *formal injunction* "not to hold, defend, teach, *and discuss in whatsoever way*" the said doctrine? If the last were true, Galileo would be guilty of defiance and serious breach of orders.

The Inquisitors produced a purported record of the formal injunction, but it was not notarized, and Galileo maintained that he had never seen it. Although there is considerable debate on this point, this document had likely been forged. Moreover, Galileo produced a letter which Cardinal Bellarmine had written to him in 1616 for use as testimony. In the letter the Cardinal had stated that he had notified Galileo that "the doctrine attributed to Copernicus . . . cannot be defended or held." The words "teach and discuss in whatsoever way" were not in this letter, nor was there mention of a formal injunction.

In these circumstances the trial turned into gradual spiritual pressure on Galileo, through appeals to his deep religiosity, to make him *believe* in a sin that he had not committed. In the end, the old fighter, who for months had been under moral torture, gave in. "I am here," he said, "to make obedience and I have not held this opinion after the decision [against it] was pronounced. . . ."

THE SENTENCE

On June 22, 1633, Galileo was taken to the large hall of a monastery in the center of Rome. There he was made to kneel on the floor while the sentence was read to him: ". . . We say, pronounce, sentence and declare that you, Galileo, for the things found in the trial and confessed by you, have made yourself . . . vehemently suspected of heresy, namely to have held and believed false doctrine, contrary to the Holy and Divine Scriptures. . . . We are agreeable that you will be absolved provided that first, with sincere heart and unfeigned faith, you abjure, curse and detest the above mentioned errors and heresies. . . . We order that the book *The Dialogue* by Galileo Galilei be prohibited by public edict. We condemn you to formal prison . . . and we impose on you as salutary penances that for the next three years you say the seven penitential psalms once a week." And Galileo, still kneeling before his judges, abjured the Copernican doctrine.

"I Galileo, son of the late Vincenzio Galilei, Florentine, aged seventy years, arraigned personally before this tribunal and kneeling before you . . . having before my eyes and touching with my hands the Holy Gospels, swear that I have always believed, do believe, and by God's help will in the future believe all that is held, preached and taught by the Holy Catholic and Apostolic Church. . . . Therefore . . . with sincere heart and unfeigned faith I abjure, curse, and detest the aforesaid errors and heresies and generally every other error, heresy, and sect whatsoever contrary to the Holy Church, and I swear that in future I will never again say or assert, verbally or in writing, anything that might furnish occasion for a similar suspicion. . . ."

According to a legend which in Italy is said to have started only in the following century, on his way back to the rooms that were his prison the old man, who had signed his sentence with shaky hands, rallied and said: *"Eppur si muove"* ("And yet it moves"). Although the words were probably not pronounced, they are significant because they show with whom popular opinion sided.

8. *Father and Daughter*

GALILEO NOW WAS AN OLD MAN, apparently defeated, for the moment certainly depressed and tired, a man looking for help and comfort. He had, as we have mentioned, three children: two girls, Virginia and Livia, and a boy, Vincenzio. Shortly after his return to Florence from Padua the two little girls had entered the old convent of San Matteo on the hill of Arcetri, in the peaceful country-side south of Florence. As soon as they had reached the age of 16, the earliest at which girls could take the vows, they had become nuns and assumed the names of Sister Maria Celeste and Sister Arcangela.

While the younger girl, Sister Arcangela, was of difficult temperament and indifferent intelligence, Sister Maria Celeste developed outstanding qualities of mind and soul. She was not only an intelligent young woman but also a very warm human being, who having given up any other love in this world. poured her affection and most tender cares on her father.

The boy Vincenzio had grown into a bright, intelligent youth but not notable for good conduct, and he had often brought worry and sorrow to his father. So Galileo did not distribute his affection equally among his three children but concentrated it on his eldest daughter.

Often, riding a small mule, he went to see his daughters in the convent, which they were not allowed to leave. He chatted with both, of course, and with other nuns, but he preferred Sister Maria Celeste's conversation to any other. When he could not make frequent visits, he and Sister Maria Celeste exchanged letters and small presents. None of the father's letters was preserved, although the daughter treasured and kept them all; it is lucky, at least, that many of hers were saved. Through her charming writing we get glimpses of Galileo's intimate life, of those unimportant acts that build up into the portrait of a man.*

GALILEO AT HOME

We see Galileo installed in a country house after he had left the city, which he blamed for his poor health. We

* Although Sister Maria Celeste's letters have more flavor in their original Italian, they are charming in English also. We recommend reading them in Mary Allan-Olney's book *The Private Life of Galileo*, London, 1870.

see him often ill despite "the good country air," but stubbornly proceeding with his studies. "It gives me some trouble," wrote Sister Maria Celeste, "to hear that Your Honor is keeping at his studies with such assiduity, because I am afraid that this will harm your health. And I should not want that, in trying to immortalize your fame, you should shorten your life, a life so much respected and held dear by us, your children, and by me in particular."

The letters picture Galileo at work early in the morning in his orchard, where he pruned olive trees and grafted grapevines, enjoying the outdoor toil that gave him the opportunity to read further in "the great book of Nature." (He was proud of being able to read this book. He said once, and repeated over and over again to his friends and children: "It cannot be understood unless one first learns to comprehend the language and read the letters in which it is composed. It is written in the language of mathematics, and its characters are triangles, circles, and other geometric figures.") "I am not a little sorry," Sister Maria Celeste wrote once, "to hear that Your Honor went back so soon to his usual exercise in the orchard because, the air being still quite raw and Your Honor weak from illness, I fear lest it might harm you." (Occasionally her advice went further and she begged him, for the sake of his health, not to overindulge, not to eat too much or drink too large quantities of wine.)

We see Galileo turn repairman for the convent and keep the nuns' clock in working shape. "I should judge," his daughter once wrote, "that the fault is in the string that, being old, does not slide easily. . . ." And a few days later, "The clock that so many times I sent back and forth now works very well." Once she asked her father to put oilcloth on the window in her cell so that she and the other nuns might have more light to do needlework. (Many windows of that time did not have glass panes, which were expensive, but only wooden shutters.) "I do not doubt of your good will, but because this is a job more befitting carpenters than philosophers, I have some fears."

In return for these small services and many others, the

nuns laundered the wide white collars and cuffs that Galileo always wore; Sister Maria Celeste cooked fruits and cakes for him and prepared simple remedies for his health.

SISTER MARIA CELESTE

In the fall of 1631 Galileo gave in to his daughter's insistence and moved to Arcetri, in a villa adjacent to the convent. The closeness reinforced the already strong intellectual and affectionate bonds between father and daughter.

It was natural that Sister Maria Celeste should be extremely sensitive to her father's satisfactions and his sorrows. She rejoiced when Cardinal Barberini became Pope, and could hardly wait for the moment when Galileo would go see this great personage. She worried during his trial, although until the very end she did not realize how serious his plight was. Galileo had remained hopeful through most of his stay in Rome, but even after he had lost hope he had hidden his feelings to spare hers.

So, at the very time when he was living the great drama of his life, he received relaxing bits of news from Arcetri in his daughter's letters. He learned that his servant needed socks, that lettuce and beans were thriving and lemons ripe, that she had tasted wine from his tuns and found it good.

Then she learned the outcome of the trial: "As much as the news of Your Honor's new affliction was sudden and unexpected, that much more my soul was pierced with extreme grief in hearing the decision finally taken both about your book and Your Honor's person. . . . My most dear father, now it is time to take advantage of that prudence that the Lord has granted you, and bear these blows with that force of soul which your religion, profession and age demand."

Galileo the Catholic, who had submitted to the will of his superiors in religion and abjured views which his rational mind knew to be right, could not have found elsewhere greater and more understanding comfort.

DEATH OF A DAUGHTER

Nature had made Galileo extremely resilient, and in a matter of days after the trial, which had left him so broken and prostrate that his friends had feared for his life, he rallied. The Holy Office mitigated the "formal prison" originally decreed and soon allowed him to go to Siena in the custody of the archbishop of that city, a learned prelate who welcomed his guest with the greatest warmth. This was all he needed to feel better. Siena is so near Florence that Sister Maria Celeste could send a trusted servant to see her father and thus be reassured about his health.

In the quiet of Siena, Galileo, in compliance with the Church's orders to give up the defense of the Copernican theory, applied himself to thinking and writing about his past studies in physics. Ironically, these very studies were the foundations of the science of mechanics, which in the hands of Newton proved the validity of the Copernican system.

After a few months in Siena, Galileo received permission to return to his villa in Arcetri, where he was to spend the rest of his life under perpetual house arrest and strict orders not to go to Florence or to let many friends visit him at one time.

Whether she knew of these restrictions or not, Sister Maria Celeste rejoiced at the news of his return. But her happiness was to be of short duration. She had always been in delicate health, and life in the convent—the scanty food, the cold, damp rooms in winter, the strict practice of religion, and the many chores performed—had undermined her resistance. A few months after her father's return she fell ill and died, at only 33 years of age.

Galileo wrote from Arcetri to a friend: "Here I was living very quietly with frequent visits to a nearby convent, where I had two daughters, both nuns, whom I loved dearly and especially the eldest, a woman of exquisite intelligence, singular goodness and very attached to me.

This daughter afflicted by melancholic humors during my absence, which she thought distressing for me, finally incurred a precipitate dysentery and died in six days, being thirty-three years of age, leaving me in extreme affliction."

9. The Last Years

ONCE MORE GALILEO REACTED TO adversity, and in his "extreme affliction" he sought comfort in work. Within two years of his daughter's death he completed *Two New Sciences,* the book to which his lasting fame as a scientist is tied, the first great work of modern physics. Like the *Dialogue, Two New Sciences* is in the form of a conversation among Salviati, Sagredo, and Simplicio, and it is divided into four days. As in the *Dialogue,* the three characters are the scientist, the well-informed layman, and the Aristotelian philosopher. In the last two days they read and comment on a Latin book of mathematical and geometric demonstrations by "Our Author," Galileo.

When Galileo tried to find a publisher for *Two New Sciences,* he came up against a new problem, a new source of grief: the Church had issued a general prohibition against printing or reprinting any of his books. But the prohibition failed to prevent publication of the book. Through a friend in Venice, Galileo's manuscript reached a very willing and able publisher, Louis Elzevir of Leyden in Holland, a Protestant country over which the Roman Church had no power. At once Elzevir undertook the printing. Galileo feigned surprise, pretended not to know how the manuscript had reached Holland. Although it is improbable that anyone believed his story, the Church let the publication of the *Two New Sciences* in 1638 go unnoticed.

Galileo did not consider his book finished. Among the works not published in his lifetime are two *Added Days,* a continuation of *Two New Sciences,* whose manuscripts are not entirely in his handwriting. He was very old then and had engaged the help of young pupils to whom he dictated his works. The pupil who most often took his dictation was that Vincenzio Viviani who many years later was to become his first and probably least accurate biographer. Viviani was 18 when he went to live in Galileo's house as his pupil, and Galileo was 75. Memory may play tricks. Perhaps the old man in telling the stories of his youth embellished them, or the young man, carried away by admiration, brought his own additions to them. Viviani himself was old (and Galileo had been dead many years) when, with unchanged devotion, he wrote the romantic story of his teacher's life.

Even before *Two New Sciences* was published, Galileo was at work on what he called a "catalogue" of the most important astronomical operations. Again, as 20 years earlier, he was trying to help navigators. "Through the quality of the instruments for optical observations and of those with which I measure times, I obtain highly accurate [results]," he wrote.

Galileo hoped to solve the problem of determining longitudes at sea. At that time navigators knew how to measure latitudes from the altitude of the pole star, but they could not determine longitudes accurately. One of their main difficulties was that of *transporting* time, of knowing while at sea what time it was at their point of departure so that they could compare this time with their local time. In short, they needed accurate clocks.

Galileo, drawing on his early researches, devised a way of measuring short intervals of time with pendulums which gave more accurate results than existing mechanical clocks. His pendulums were manually operated and would stop unless someone gave them "a lively push" when needed. In trying to improve this primitive method of keeping pendulums going, he invented an escapement (the mechanism through which the energy of a weight is delivered to the pendulum) and came very close to building the first pendulum clock. But blindness intervened, then death. All he could do was to describe the

Galileo's Design for a Clock. Galileo was blind at this time, and the drawing was made on his instructions by his son. It shows the pendulum and the escapement.

instrument he had in mind to his son, so accurately that Vincenzio could make a blueprint from which two centuries later a working pendulum clock was built.

The glory of actually building the first pendulum clock was left to the Dutch physicist and astronomer Christian Huygens. He described his clock in a book, *Horologium Oscillatorium,* and in the preface he acknowledged Galileo's part in the invention.

BLINDNESS

Galileo's eyes had bothered him for many years. He first complained about them after he had finished the *Dialogue,* believing that long hours of reading and correcting proofs were the cause of his trouble. On the evidence of his symptoms as he described them in his letters, modern physicians have diagnosed his eye illness as glaucoma. He lost the sight of his right eye first, then of the left. During the last four years of his life he was completely blind.

He informed a friend of his misfortune in these terms: "Your dear friend and servant Galileo is irreparably and completely blind; in such a way that that sky, that world and that universe, which with my wondrous observations and clear demonstrations I amplified a hundred and thousand times over what was believed most commonly by the learned of all past centuries, is for me now so diminished and narrowed that it is no greater than what my body occupies."

Several years earlier, when still far from nostalgic contemplation of his past and led by his precise scientific consciousness, he had written: "As I am considering the world which is perceived by our senses I cannot say in absolute manner whether it is large or small: I may well say that it is very large by comparison with the world of worms, who, having no other means to measure it than the sense of feeling, cannot judge it [to be] larger than that space which they occupy. And it is not repugnant to me that the world perceived through our senses may be as small in comparison with the universe as the world of the worms by respect to ours. Concerning then what the intellect, beyond the senses, might apprehend, my mind cannot adapt itself to conceive it either finite or infinite. . . ."

MILTON'S VISIT

Galieo was already blind when a young, little-known English poet visited him in Arcetri. John Milton, then 29 years old, was on a tour of Europe to enlarge his already extraordinary culture, to talk with the learned and observe their ways. Of his visit to Galileo little is known, but the context in which Milton left its only record is highly significant. In his famous *Areopagitica,* a speech addressed to Parliament against an ordinance requiring the licensing of all books, Milton wrote:

"I could recount what I have seen and heard in other countries, where this kind of Inquisition tyrannizes . . . that this was it which had damped the glory of Italian wits, that nothing had been there written now these many years but flattery and fustian. There it was that I found and visited the famous Galileo, grown old, a prisoner of the Inquisition, for thinking in astronomy otherwise than the Franciscan and Dominican licensers thought."

Milton was right. In Italy as in other Catholic countries Galileo's trial and the prohibition of his books had the effect of temporarily halting the development of the science of cosmology. Even the French mathematician René Descartes (1596-1650) thought it best to keep quiet for a while, as he said in a letter to his and Galileo's friend, the French Father Marin Mersenne. The boldest thinkers had keenly followed Galileo's vicissitudes and were looking at the future with misgivings. But time was on Galileo's side. In 1635 a Latin translation of the *Dialogue* was published in Protestant Strasbourg. Science evolved at its own sure pace, and eventually the Church adopted the "thinking in astronomy otherwise than the Franciscan and Dominican licensers thought." In 1757 Pope Benedict XIV repealed the general prohibition against books which taught that the sun is stationary and the earth revolves. In 1822 Pope Pius VII approved the

printing of books that treated the motion of the earth
not as a mere mathematical hypothesis but as established
truth, "in accordance with the opinion of modern astronomers." And finally, in the catalogue of prohibited
books of 1835 neither Copernicus's *De Revolutionibus* nor
Galileo's *Dialogue* was any longer listed.

DEATH

In November 1641 a slow fever seized Galileo and
his arthritic pains became more acute. This was his last
illness. He died on January 8, 1642, at almost 78 years
of age.

Florence voted to erect a monumental tomb to him in
the Church of Santa Croce, where great Florentines are
buried. But Urban VIII, whose rancor had not subsided,
who had repeatedly refused Galileo's petitions for more
freedom during the years of "prison" in Arcetri, forbade
even this last honor. Galileo's body lay for almost a century in the basement of the church. But in the end his remains found a worthy resting place inside the church, in
a marble sarcophagus at the center of a large monument
erected in accordance with the last will of his devoted
pupil Vincenzio Viviani.

10. Galileo's Physics

THE FULL TITLE OF GALILEO'S greatest scientific book,
which we now call simply *Two New Sciences,* was *Dis-
courses and Mathematical Demonstrations concerning
Two New Sciences Pertaining to Mechanics and Local
Motions.* It was not Galileo's original title, and he protested
to the Dutch publisher, L. Elzevir, for the liberty he had
taken in changing it. But despite Galileo's dislike, the title
could not be more meaningful. The book is a healthy em-
bryo of a new science, the first chapter of a "natural

philosophy" or "physics" which in the next three centuries spread from "local motions" to all motion in the universe, and from "mechanics" to optics, electricity, and all the other chapters into which a modern physics textbook is divided.

The "newness" of Galileo's book did not consist in the subjects treated but in the way in which they were treated. Most of the subjects were old. Part of the "mechanics" summarized and developed the content of a treatise, *The Mechanics,* which Galileo had written in Padua, and it dealt with traditional problems of statics and simple machines. "Local motions" were, as Galileo himself remarked, "a very ancient subject," since "in nature perhaps nothing is older than motion." Other parts were new and included an important contribution to the study of the resistance of materials.

The fact that the subjects were limited and traditional excluded the possibility of great fundamental discoveries. The study of how far a beam fastened to a wall would resist stress without breaking or of the motion of projectiles held no sensational finding in store. Thousands of years before Galileo, men had used beams to build temples, and David's sling was certainly not the first to throw a stone. But beams and moving stones had never before been the object of so profound and successful a scientific investigation, which required Galileo's own way of thinking and his philosophy. Thus the "new" in the *Two New Sciences* consisted in the scientific approach to old subjects.

Modern society has developed on the basis of this great innovation in the way of looking at the world around us. Scientific investigation is so much a part of our culture that we are no longer aware of it. Yet we daily apply its methods and results, as much in the field of industry and production as in private life—in prospecting for oil and gold as in canning food, in planning street lighting as in using a light meter to determine the camera aperture size for taking a picture.

This type of scientific investigation is often called the "experimental method." We have already seen an example

of it in Galileo's study of pendulums, and we shall soon
see other applications. In stressing and practicing this
method Galileo was a pioneer, for his way of doing
science had no large following among his contemporaries,
not even among the greatest of them, such as Descartes
and Kepler.

THE LAW OF INERTIA

Perhaps the most gratifying aspect of culture, an aspect
that gives us a measure of the value of human dignity,
is the possibility of reconstructing in ourselves the spiritual
processes of our great predecessors. For this reason we
may find it more interesting, rather than examine the content of the *Two New Sciences,* to try to reconstruct the
long and straight path that Galileo followed in going from
his early Aristotelian studies to his last, greatest work.

In this long path we often find reasons for surprise.
It seems almost incredible that *De Motu* and *Two New
Sciences* can be the works of the same person. We must
keep in mind that one was the product of his youth and
the other of his maturity.

When Galileo wrote *De Motu,* while lecturing in Pisa,
he was much more under the influence of traditional
thinking that he himself realized. The points brought up in
De Motu are an indication of this influence. In Chapter 5,
for instance, Galileo discussed "Whether bodies in *their
proper* places are heavy or light." This idea of a place
proper to each body and determined by its nature—
namely, by the matter of which it is formed—was connected with Aristotle's idea of four essential elements and
of their distribution in the universe.

Other questions discussed in *De Motu* were: (1)
What causes "natural" motion, and (2) What causes
"violent" motion. Galileo explained that "natural motion
is that by which bodies, in proceeding, get closer to their
proper places, violent motion that of bodies which in
moving get away from these [places]." The proper
place of bodies was the center of the earth and only
motion downward was "natural." Thus if a body fell freely

along an inclined plane, it moved with "natural" motion, while if it was thrown and made to go up the inclined plane it moved with "violent" motion.

These ideas of natural and violent motion were Aristotelian. But in contrast with Aristotelian physics, in which fire naturally rose, Galileo explains at length that all bodies are heavy, so that even fire would move downward if the air were removed. He also added a motion neither violent nor natural—that on a horizontal plane. He stated that "on a smooth horizontal plane any body, meeting with no external resistance, can be moved by as small a force as wished." A question may arise at this point: How would this body, whose motion is not opposed by external resistance, behave if no force at all acted upon it?

Young Galileo probably asked himself this question, but at that time he did not find an answer. In *De Motu,* as we have seen, he refuted the opinion that the motion of a body is maintained by the air around it, and he accused Aristotle and his followers of having fallen into this error because they were incapable of conceiving that a body can be moved by "impetus" (*virtù impressa*) or of understanding what impetus was.*

In Galileo's mind, however, the thought persisted that the problem of a motion neither natural nor violent, such as motion on a horizontal plane, hid something of basic importance. Later, insistently, in almost all his works he hinted at partial or tentative answers to this question, and as time went on this thought became clearer and clearer. Finally, in *Two New Sciences,* more than 40 years after *De Motu,* he formulated his answer. In the third day, talking of a moving body, he said:

". . . along a horizontal plane the motion is uniform

---

* In the centuries before Galileo, the commentators on Aristotle had introduced the concept of impetus. Because it sometimes played the role of momentum and sometimes that now played by kinetic energy, this concept was confused. Not even Galileo fully understood it. The two ideas in it were clearly separated only after Galileo.

since here it experiences neither acceleration nor retardation . . ." and ". . . any velocity once imparted to a moving body will be rigidly maintained as long as the external
causes of acceleration or retardation are removed, a condition which is [experimentally] found only on horizontal planes. . . ." This velocity, he added, if acting
alone "would carry the body at a uniform rate to in-
finity. . . ."

This is a correct formulation of the "First Law of
Motion," or "Law of Inertia," which Newton expressed
in its definitive form a few decades later. Newton himself said that he had learned the law of inertia from Galileo.

Galileo took so long to formulate this law for two
main reasons. First, it required a complete break from
the Aristotelian school, which, among others, considered
motion inconceivable in the absence of a force. Second,
and perhaps more important, it needed a process of abstraction and rational generalization entirely new in the
study of natural phenomena. This process is constantly
repeated throughout the *Two New Sciences.*

In formulating the "Law of Inertia" the abstraction
consisted of imagining the motion of a body on which
no force was acting and which, in particular, would be
free of any sort of friction. This abstraction was not
easy, because it was friction itself that for thousands of
years had kept hidden the simplicity and validity of the
laws of motion. In other words, friction is an essential
element in all human experience; our intuition is domi-
nated by friction; men can move around because of friction; because of friction they can grasp objects with their
hands, they can weave fabrics, build cars, houses, etc.
To see the essence of motion beyond the complications
of friction indeed required a great insight.

As we have said, tentative formulations of the First
Law of Motion appeared several times in Galileo's works,
but in most they were not so clear and explicit as the
passage we have just quoted. For a long time the first
law was somewhat confused in his mind with the idea

that a horizontal plane is a portion of a huge plane that follows the horizon and bends around the earth. According to this idea, motion on a horizontal plane is circular motion along a circle of huge radius.

Although Galileo had long since freed himself from most of Aristotle's wrong opinions, for many years he could not entirely discard the mystical idea, still current in his day, of the "perfection" of circular motion. Mysticism often pervaded the thinking of great men of science. A century after Galileo's death, the French mathematician Pierre Maupertuis was still looking for the "metaphysical reason of the Creator's preference for the laws of the inverse proportion of the squares of the distances in universal attraction." Galileo also was in a sense a mystic, in his deep, serene religiosity, in his devout admiration of the Creator and the life He created, of nature in its every aspect. He took a long time to discard ideas such as the perfection of uniform circular motion.

ABOUT FALLING BODIES

The third and fourth days of *Two New Sciences* concern uniformly accelerated motion and motion of projectiles, respectively. They contain the great discoveries and innovations to which Galileo's lasting scientific fame is tied. They put into evidence and systematically repeat the essential steps of Galileo's thinking process. We have seen these steps already in his study of pendulums and less explicitly in the formulation of the first law of motion. We may retrace them again, more carefully and extensively, in the case of uniformly accelerated motion, which Galileo studied mostly as a particular instance of freely falling bodies. The best way of doing this is to examine what Galileo said in the third day of *Two New Sciences*, through the mouths of Salviati, Sagredo, and Simplicio.

Salviati begins by saying that to give free vent to imagination is quite different from describing a phenomenon correctly and objectively: ". . . anyone may invent an arbitrary type of motion and discuss its properties ... but we have decided to consider the phenomena of bodies falling with an acceleration such as actually occurs in nature and to make this definition of accelerated motion exhibit the essential features of *observed* accelerated motions." (Translators' italics.)

Observation indicates that a falling stone acquires an always increasing velocity, and Galileo asks himself: "Why should I not believe that such increases take place in a manner which is exceedingly simple . . . ?" Then he advances a theoretical hypothesis based on this criterion of simplicity: "We find . . . no increment more simple than that which repeats itself always in the same manner. . . . thus we may picture to our mind a motion as uniformly and continuously accelerated when, during whatever equal intervals of time, equal increments of speed are given to it. . . . And thus . . . *we put the increment of speed as proportional to the increment of time.*" (Editors' italics.)

This type of motion called for the introduction of a difficult concept: a speed that varied continuously from instant to instant and that, therefore, did not correspond to the simple ratio of space traveled to time needed to travel it. Salviati gives an example: "To put the matter more clearly, if a body were to continue its motion with the same speed which it had acquired during the first time-interval and were to retain this same uniform speed, then its motion would be twice as slow as that which it would have if its velocity had been acquired during *two* time-intervals."

And later on, since Sagredo and Simplicio keep raising difficulties, he adds that "each time-interval however small may be divided into an infinite number of instants" and these "correspond to the infinite degrees [values] of velocity."

Having solved the problem of explaining instant velocity, Galileo must verify his theoretical scheme. He had assumed, as we have seen, that in uniformly accelerated motion "the increment of speed [that is, acceleration] is proportional to the increment of time." To verify this would be a simple problem if he could measure times and speeds easily. But he faces two difficulties: first, the speeds that freely falling bodies acquire soon become very great and impractical to measure; second, and more important, he wants to measure instantaneous velocities, and to do this he either has to find a way of letting the falling bodies "continue their motion with the same speed" or must discover a less direct but equally convincing method of proving that his assumption is right. Mathematics provides this equally convincing method. He proves that if his assumption is true, then "the spaces described by a body falling from rest with a uniformly accelerated motion are to each other as the squares of the time-intervals employed in traversing these distances."

Simplicio, however, finds mathematical demonstrations (implying elements of calculus) rather obscure. He is not convinced that Salviati's description of uniformly accelerated motion holds also for falling bodies in nature, and asks him to describe an experiment.

Salviati does this, explicitly saying that the experiment is Galileo's. By an ingenious expedient Galileo had replaced the too-fast vertical fall with fall along an inclined plane, showing that the latter reproduces the former, as if seen, we should now say, in slow motion.

Salviati says: "A piece of wooden moulding or scantling, about 12 cubits long, half a cubit wide, and three finger-breadths thick, was taken; on its edge was cut a channel a little more than one finger in breadth; having made this groove very straight, smooth, and polished, and having lined it with parchment, also as smooth and polished as possible, we rolled along it a hard, smooth, and very round bronze ball. Having placed this board in a sloping position, by lifting one end some one or two cubits above the other, we rolled the ball, as I was just saying, along the channel, noting, in a manner presently to be described, the time required to make the descent. We repeated this experiment more than once in order to measure the time with an accuracy such that the deviation between two observations never exceeded one-tenth of a pulse-beat.

"Having performed this operation and having assured ourselves of its reliability, we now rolled the ball only one-quarter the length of the channel; and having measured the time of its descent, we found it precisely one-half of the former. Next we tried other distances comparing the time for the whole length with that for the half, or with that for two-thirds, or three-fourths, or indeed for any fraction; in such experiments, repeated a full hundred times, we always found that the spaces traversed were to each other as the squares of the times, and this was true for all inclinations of the plane; that is, of the channel, along which we rolled the ball. We also observed that the times of descent, for various inclinations of the plane, bore to one another precisely that ratio which, as we shall see later, the Author [Galileo] had predicted and demonstrated for them.

"For the measurement of time, we employed a large vessel of water placed in an elevated position; to the bottom of this vessel was soldered a pipe of small diameter giving a thin jet of water, which we collected in a small glass during the time of each descent, whether for the whole length of the channel or for a part of its length; the water thus collected was weighed, after each descent, on a very accurate balance; the differences and ratios of these weights gave us the differences and ratios of the times, and this with such accuracy that although the operation was repeated many, many times, there was no appreciable discrepancy in the results."

Galileo's ingenuity is striking. As most clocks of the time are not accurate, but scales are, he turns the measure of times into measures of weights. Yet more important than ingenuity, more essential to modern science, is the replacing of the phenomenon as it happens in nature with an equivalent phenomenon especially *devised* and *produced at will*, under much simpler conditions: Friction is here reduced to a minimum, and times are long enough to be accurately measured.

Galileo's experiment of the inclined plane has been repeated several times in recent years. Although the results in these tests are not as accurate as Galileo claimed, they are adequate to confirm his theoretical prediction.

THE MOTION OF PROJECTILES

In the fourth day of *Two New Sciences,* Galileo deals with the motion of projectiles and makes considerations of fundamental importance for the interpretation of experiments in general.

First of all, he introduces a new principle, now considered fundamental in mechanics, the principle of independent simultaneous motions. A body can move as if with two different and independent motions at the same time, and the path that it describes can be obtained by combining the two paths that the body would describe if it moved with one motion at a time.

Galileo explains this principle in these terms: "I now propose to set forth those properties which belong to a body whose motion is compounded of two other motions, one uniform and one naturally accelerated. . . . This is the kind of motion seen in a moving projectile; its origin I conceive to be as follows:

"Imagine any particle projected along a horizontal plane without friction; then we know . . . that this particle will move along this same plane with a motion which is uniform and perpetual, *provided the plane has no limits.* But if the plane is limited and elevated, then the moving particle will . . . acquire, in *addition* to its previous uniform and perpetual motion, a downward propensity due to its own weight." (Editors' italics.)

He then shows, through a geometrical construction, that "a projectile which is carried by a uniform horizontal motion compounded with a naturally accelerated vertical motion describes a path which is a semi-parabola."

Both Sagredo and Simplicio raise objections. Sagredo finds it difficult to accept the fact that a projectile moves along a parabola. He says: "One cannot deny that the argument is new, subtle and conclusive, resting as it does upon this hypothesis, namely that the horizontal motion

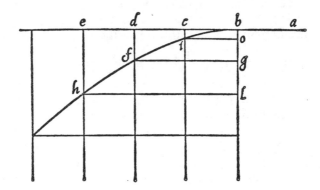

Galileo's Diagram shows that the path of a projectile is a parabola. If there were no gravity, the projectile would traverse the equal distances *bc, cd, de* in equal intervals of time. But at *b* gravity starts to pull the projectile downward and makes it fall the distances *bo, og, gl,* proportional to the squares of the times. So at the end of the first interval of time the projectile has traveled a horizontal distance *bc* and fallen a vertical distance *bo,* and it is at *i.* At the end of the second interval of time it is at *f,* at the end of the next at *h,* and so on. The line *bifh* is a parabola.

remains uniform, that the vertical motion continues to be [uniformly] accelerated downwards . . . and that such motions . . . as these combine without altering, disturbing, or hindering each other, so that as the motion proceeds the path of the projectile does not change into a different curve: but this in my opinion is impossible." Sagredo's reason for his opinion is that if the projectile starting its motion from *b* actually followed the parabolic path *bifh,* it would move farther and farther from the vertical *bn.* This vertical ends at the center of the earth. The projectile, then, would get farther and farther away from the center of the earth and could never reach it. This Sagredo finds hard to believe.

Simplicio finds a different difficulty and says, "I do not see how it is possible to avoid the resistance of the medium which must destroy the uniformity of the horizontal motion and change the law of acceleration of falling bodies."

Salviati answers, admitting that both Sagredo's and Simplicio's objections are valid: "I grant that these conclusions proved in the abstract will be different when applied in the concrete and . . . neither will the horizontal motion be uniform nor the natural acceleration be in the ratio assumed, nor the path of the projectile a parabola. . . ."

This path would be "a true parabola if the distance from the center of the earth were infinite." But "our distance from the center of the earth is not really infinite, but merely very great in comparison with the small dimension of our apparatus. . . . The range of our projectiles . . . will never exceed four of those miles of which as many as a thousand separate us from the center of the earth; and since these paths terminate upon the surface of the earth only very slight changes can take place in their parabolic figure which, it is conceded, would be greatly altered if they terminated at the center of the earth." Later he says explicitly: "We are able . . . to say that *the errors* . . . in the results . . . *are small* in the case . . . where . . . the distances are negligible in comparison with the semi-diameter of the earth. . . ." (Editors' italics.)

Answering Simplicio, Salviati then says: "As to the perturbation arising from the resistance of the medium, this is more considerable and does not, on account of its manifold forms, submit to fixed laws and exact description . . . the greater [the velocity] is, the greater will be the resistance offered by the air; a resistance which will be greater as the moving bodies become less dense. So . . . no matter how heavy the body, if it falls from a considerable height, the resistance of the air will be such as to prevent any increase in speed and will render the motion uniform; and in proportion as the moving body is less dense this uniformity will be so much the more quickly attained. . . .

"Even horizontal motion which, if no impediment were offered, would be uniform and constant is altered by the resistance of the air and finally ceases; and here again the less dense the body the quicker the process. Of these properties of weight, of velocity, and also of form, infinite in number, it is not possible to give any exact description; hence in order to handle this matter in a scientific way, it is necessary to cut loose from these difficulties; and having discovered and demonstrated the theorems, in the case of no resistance, to use them and apply them *with such limitations as experience will teach.*" (Editors' italics.)

In Salviati's answers to Sagredo's and Simplicio's well-founded objections, Galileo illustrates another essential element of his approach to scientific research. In setting up his experiments and interpreting the results, he carefully estimates *experimental errors* and theoretical *approximations*. To Galileo experiments meant measurements, for in his maturity he considered proper subjects of science only those which are susceptible of measurement. He knew that if he repeated his experiments time after time he should always expect small, accidental, irregular deviations of his measures from their average values, and that these deviations were unavoidable.

It is not equally simple to evaluate the limit to which a theoretical scheme may push the simplification of a phenomenon and still be valid. The idealized phenomenon in the theoretical scheme may find experimental confirmation *only* if the simplification involves *systematic* (that is, expected and understood) deviations from the results of observation smaller than the *accidental* deviations.

Salviati says in essence that they have replaced the large variety of actual phenomena, which are influenced by most varied "accidents" of form, of weight, etc., with a *single* ideal, extremely simple phenomenon. In their experiment they tried to *approximate* as much as possible the simple conditions of an ideal phenomenon, but they *know* that they can do this only to a certain point. They know, for instance, that it will be practically impossible to observe the departure of the true path of a projectile from a parabola *provided* the heights from which projectiles are thrown are very small in respect to the radius of the earth. Under these conditions the effect of the curvature of the earth and the variations of the acceleration of gravity with the distance from the center are negligible, and the direction of this acceleration is constant. Hence, in "local motion" the acceleration of gravity is everywhere constant in magnitude and direction.

In a similar way Salviati says that they know the resistance of air will not appreciably alter the velocity of a thrown body *provided* this velocity is not too great (that is, very small in comparison with the constant velocity that the body would acquire after a long fall through the air).

Galileo's great contemporary, Descartes, writing about *Two New Sciences* a few years after its publication, maintained that he had nothing to learn from Galileo and that Galileo's theory was wrong, leading to results that went against those of observation. Descartes, too exclusively a mathematician and rational philosopher, had not understood that Galileo's theory was valid, as any other physical theory is valid, "with such limitations as experience will teach us."

FAITH IN THE LAWS OF NATURE

In studying the motion of projectiles and falling bodies Galileo continuously alternated theoretical considerations and actual experiments, and compared quantitative results of experiments with conclusions obtained through theorizing, making allowance for experimental errors.

In this way of proceeding, Galileo placed an absolute faith in the fact that nature cannot be deceitful; that between the mathematical, rational, and objective interpretation of phenomena and their actual happening and repetition there can be no unpleasant surprise. Each experiment, if sensibly conceived, is a date with nature to which nature responds faithfully, at the exact time, without cheating or misleading.

Experiments can, if properly carried out and interpreted, establish a "law" of nature. We can then foresee the results of numberless other experiments that fit into the pattern of that law without actually performing them. "The knowledge of a single fact acquired through a discovery of its causes prepares the mind to understand and ascertain other facts without need of recourse to experiment."

To Galileo the belief in the existence of natural laws so generally valid and reproducible that they rule the universe was equivalent to a faith, based on the conviction that the proper interpretation of natural phenomena cannot contradict the laws of human reason. He expressed this beautifully in the already quoted passage of the letter to Cristina which starts: "Nature . . . is inexorable and immutable."

"THIS VAST AND MOST EXCELLENT SCIENCE"

It is certain, in our opinion, that Galileo performed a large number of experiments, and we probably do not know them all. A few, such as his first with pendulums or that of weights dropped from the Leaning Tower of Pisa, were related by his pupils and became confused in legend. Often Galileo himself described them only sketchily, insisting on their meaning without indicating the limits of errors and therefore their accuracy. Here and there his extreme ingenuity in conceiving experiments flared up, as, for instance, when he tried to measure the velocity of light in order to ascertain whether its propagation is instantaneous. He set up the experiment in this manner: Two men stood at a distance from each other, each holding a light that he could cover and uncover. One of the men uncovered his light, and as soon as the other saw it he uncovered *his*. Thus the two were sending signals to one another. The distance between the two men was then increased to see whether this increase caused any delay between the signals. A delay, Galileo said, would prove that the propagation of light is not instantaneous. The experiment failed, of course, and he concluded correctly: "In fact I have tried the experiment only at a short distance, less than a mile, from which I have not been able to determine with certainty whether the appearance of the opposite light was instantaneous or not; but if not instantaneous it is extraordinarily rapid."

The experiment failed because men's reactions (here the uncovering of the lamp) take much longer than it takes light to travel the distance of a mile (about eighteen millionths of a second).

About fifty years later, in 1675, the Danish astronomer Ole Roemer (1644-1710) had the genial idea of using astronomical rather than terrestrial distances. His "lamp" was one of Jupiter's satellites, "covered" or "uncovered" by its own eclipses; and the distance traveled by light was that of the earth's orbit.

Despite his brilliance in devising and performing experiments, Galileo's main contribution to physics and to science as a whole was not a great discovery comparable, for example, to Newton's universal gravitation. Rather, he paved the way for many discoveries. Max Born, for instance, has said: "The inner logic of Galilean mechanics was so strong that Newton was able to take the great step of applying it to the motion of the stars."

Galileo's main contribution to science was, so to speak, that of a great teacher. He taught that mathematics, measurement, and the reproduction of phenomena under controlled and simplified conditions are essential to the study of nature. In doing this he made a final break from the Aristotelian tradition and led physics on a new path from which there has been no deviation. Physicists in our day are still following the example he set on the ways of searching for scientific truth. His ability to use and refine common experience, his habit of rigorous, logical thinking, the consistent use of stringent "thought" experiments, are now essential parts of the training and practice of scientists. His first heirs in this tradition were Huygens, Torricelli, Pascal, Mariotte, Boyle, and Newton. From these few, in little over three centuries, Galileo's influence spread to all men capable of receiving a scientific education. His influence extended beyond the field of pure science, for he was also a master at controversy and popularization. Here he stands out for his desire to share his knowledge with others, for presenting it in an enjoyable form, tempering wit and argument with the urbanity of the Venetian noblemen and Medicean Court.

We may well believe that he was conscious of his role in science and that a deep emotion inspired him when in his clear and lively language he wrote, as an introduction to the third day of *Two New Sciences*, what we may consider his own summary of all we have said about him.

"My purpose is to set forth a very new science dealing with a very ancient subject. There is, in nature, perhaps nothing older than motion, concerning which the books written by philosophers are neither few nor small; nevertheless I have discovered by experiment some properties of it which are worth knowing and which have not hitherto been either observed or demonstrated. Some superficial observations have been made, as, for instance, that the natural motion of a heavy falling body is continuously accelerated; but to just what extent this acceleration occurs has not yet been announced. . . .

"It has been observed that missiles and projectiles describe a curved path of some sort; however, no one has pointed out the fact that this path is a parabola. But this and other facts, not few in number or less worth knowing, I have succeeded in proving; and what I consider more important, there have been opened up to this vast and most excellent science, of which my work is merely the beginning, ways and means by which other minds more acute than mine will explore its most remote corners."

APPENDIX

The Little Balance

GALILEO WAS 22 years old when he wrote *La Bilancetta* (*The Little Balance*), his first work in Italian. This piece of work reveals both his unlimited admiration for Archimedes and the fact that his ingenuity and craftsmanship were well developed at an early age. So far as writing is concerned, it is not too good. Young Galileo's style is intricate; his sentences wind about; his use of words and grammar is not always precise.

In translating this piece we have tried to save at least some of the original flavor and imprecision in order to provide a comparison with Galileo's later writings. We wish to note, for instance, that the Italian word *misto,* by which Galileo indicated the end product obtained by mixing two metals, should be rendered in modern English as "alloy." Galileo does not use, however, the Italian word for alloy, which is *lega* and which was certainly in use in his time. Accordingly, we have translated *misto* into "mixture."

Some confusion arises from Galileo's use of the word *gravità* or "gravity." As in Old English gravity meant weight, so it did in old Italian. But scientific terminology was not yet well established, and Galileo uses *gravità* both for "weight" and for "density" or "specific gravity." (Mass and weight were not yet scientifically defined as two different concepts, and both "density" and "specific gravity" might be used to render *gravità* in the second context of Galileo's use of it.) For the sake of clarity, we have translated *gravità* into either "weight" or "specific gravity" according to the context.

THE LITTLE BALANCE

Just as it is well known to anyone who takes the care to read ancient authors that Archimedes discovered the jeweler's theft in Hiero's crown, it seems to me the method which this great man must have followed in this discovery has up to now remained unknown. Some authors have written that he proceeded by immersing the crown in water, having previously and separately immersed equal amounts [in weight] of very pure gold and of silver, and, from the differences in their making the water rise or spill over, he came to recognize the mixture of gold and silver of which the crown was made. But this seems, so to say, a crude thing, far from scientific precision; and it will seem even more so to those who have read and understood the very subtle inventions of this divine man in his own writings; from which one most clearly realizes how inferior all other minds are to Archimedes's and what small hope is left to anyone of ever discovering things similar to his [discoveries]. I may well believe that, a rumor having spread that Archimedes had discovered the said theft by means of water, some author of that time may have then left a written record of this fact; and that the same [author], in order to add something to the little that he had heard, may have said that Archimedes used the water in that way which was later universally believed. But my knowing that this way was altogether false and lacking that precision which is needed in mathematical questions made me think several times how, by means of water, one could exactly determine the mixture of two metals. And at last, after having carefully gone over all that Archimedes demonstrates in his books *On Floating Bodies* and *Equilibrium,* a method came to my mind which very accurately solves our problem. I think it probable that this method is the same that Archimedes followed, since, besides being very accurate, it is based on demonstrations found by Archimedes himself.

This method consists in using a balance whose construction and use we shall presently explain, after having expounded what is needed to understand it. One must first know that solid bodies that sink in water weigh in water so much less than in air as is the weight in air of a volume of water equal to that of the body. This [principle] was demonstrated by Archimedes, but because his demonstration is very laborious I shall leave it aside, so as not to take too much time, and I shall demonstrate it by other means. Let us suppose, for instance, that a gold ball is immersed in water. If the ball were made of water it would have no weight at all because water inside water neither rises nor sinks. It is then clear that in water our gold ball weighs the amount by which the weight of the gold [in air] is greater than in water. The same can be said of other metals. And because metals are of different [specific] gravity, their weight in water will decrease in different proportions. Let us assume, for instance, that gold weighs twenty times as much as water; it is evident from what we said that gold will weigh less in water than in air by a twentieth of its total weight [in air]. Let us now suppose that silver, which is less heavy than gold, weighs twelve times as much as water; * if silver is weighed in water its weight will decrease by a twelfth. Thus the weight of gold in water decreases less than that of silver, since the first decreases by a twentieth, the second by a twelfth.

Let us suspend a [piece of] metal on [one arm of] a scale of great precision, and on the other arm a counterpoise weighing as much as the piece of metal in air. If we now immerse the metal in water and leave the counterpoise in air, we must bring the said counterpoise closer to the point of suspension [of the balance beam] in order to balance the metal. Let, for instance, ab be the balance [beam] and c its point of suspension; let a piece of some metal be suspended at b and counterbalanced by the weight d. If we immerse the weight b in water the weight d at a in the air will

* Modern values for the specific gravities of pure gold and silver are 19.3 and 10.5, respectively.

weigh more [than *b* in water], and to make it weigh
the same we should bring it closer to the point of suspension *c,* for instance to *e.* As many times as the distance *ac* will be greater than the distance *ae,* that many
times will the metal weigh more than water.

Let us then assume that weight *b* is gold and that
when this is weighed in water, the counterpoise goes
back to *e;* then we do the same with very pure silver
and when we weigh it in water its counterpoise goes in *f.*
This point will be closer to *c* [than is *e*], as the experiment shows us, because silver is lighter than gold.
The difference between the distance *af* and the distance
ae will be the same as the difference between the
[specific] gravity of gold and that of silver. But if we
shall have a mixture of gold and silver it is clear that
because this mixture is in part silver it will weigh less
than pure gold, and because it is in part gold it will
weigh more than pure silver. If therefore we weigh it in
air first, and if then we want the same counterpoise to
balance it when immersed in water, we shall have to
shift said counterpoise closer to the point of suspension
c than the point *e,* which is the mark for gold, and
farther than *f,* which is the mark for pure silver, and
therefore it will fall between the marks *e* and *f.* From the
proportion in which the distance *ef* will be divided we
shall accurately obtain the proportion of the two metals
composing the mixture. So, for instance, let us assume
that the mixture of gold and silver is at *b,* balanced in
air by *d,* and that this counterweight goes to *g* when the
mixture is immersed in water. I now say that the gold
and silver that compose the mixture are in the same proportion as the distances *fg* and *ge.* We must however
note that the distance *gf,* ending in the mark for silver,
will show the amount of gold, and the distance *ge* ending in the mark for gold will indicate the quantity of
silver; so that, if *fg* will be twice *ge,* the said mixture will
be of two [parts] of gold and one of silver. And thus,
proceeding in this same order in the analysis of other
mixtures, we shall accurately determine the quantities of
the [component] simple metals.

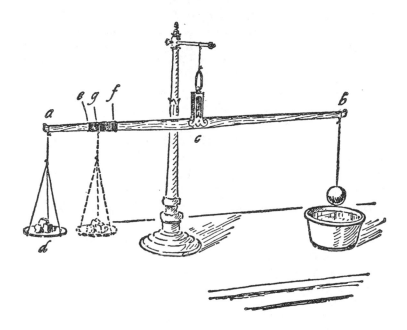

The Little Balance. A drawing based on Galileo's description.

To construct this balance, take a [wooden] bar at least two *braccia* * long—the longer the bar, the more accurate the instrument. Suspend it in its middle point; then adjust the arms so that they are in equilibrium, by thinning out whichever happens to be heavier; and on one of the arms mark the points where the counterpoises of the pure metals go when these are weighed in water, being careful to weigh the purest metals that can be found. Having done this, we must still find a way by which easily to obtain the proportions in which the distances between the marks for the pure metals are divided by the marks for the mixtures. This, in my opinion, may be achieved in the following way.

On the marks for the pure metals wind a single turn

---

* The Florentine *braccio* (plural *braccia*) was equal to 58.4 cm., or approximately 22½ inches.

of very fine steel wire, and around the intervals between marks wind a brass wire, also very fine: these distances will be divided in many very small parts. Thus, for instance, on the marks *e, f* I wind only two turns of steel wire (and I do this to distinguish them from brass); and then I go on filling up the entire space between *e* and *f* by winding on it a very fine brass wire, which will divide the space *ef* into many small equal parts. When then I shall want to know the proportion between *fg* and *ge* I shall count the number of turns in *fg* and the number of turns in *ge,* and if I shall find, for instance, that the turns in *fg* are 40 and the turns in *ge* 21, I shall say that in the mixture there are 40 parts of gold and 21 of silver.

Here we must warn that a difficulty in counting arises: Since the wires are very fine, as is needed for precision, it is not possible to count them visually, because the eye is dazzled by such small spaces. To count them easily, therefore, take a most sharp stiletto and pass it slowly over said wires. Thus, partly through our hearing, partly through our hand feeling an obstacle at each turn of wire, we shall easily count said turns. And from their number, as I said before, we shall obtain the precise quantity of pure metals of which the mixture is composed. Note, however, that these metals are in inverse proportion to the distances: Thus, for instance, in a mixture of gold and silver the coils toward the mark for silver will give the quantity of gold, and the coils toward the mark for gold will indicate the quantity of silver; and the same is valid for other mixtures.

A Note on "The Little Balance"

Although it is true, as Galileo says, that Archimedes could have made a calculation of the composition of the crown, there is no evidence that he actually did so. The crown anecdote was first told in the first century B.C. by the famous Roman architect Vitruvius, who described only the simple proof (by observation of the dissimilarity between the volumes of water caused to overflow by the crown and by an equal weight of gold) that the crown was made of adulterated gold.

The use of density as an index of the composition of an alloy involves the assumption that the volumes of the two metals remain unchanged on mixing with each other —an assumption which most early writers made without realizing its significance. It is not exactly true, for atoms may pack differently with their own kind from the way they pack with those of another element. The chemist Glauber was the first, in 1648, to look into this by experiment, and he obtained very uncertain results, because of the variable amount of porosity in his castings.

The general principles of assay by density were well known long before Galileo's time. The first record of an actual determination of composition by density measurement is in a Latin poem, *Carmen de ponderibus,* which was printed in 1475 but which had circulated in manuscript copies for centuries before: it was probably written by the Latin grammarian Priscianus, who flourished about A.D. 500. The poem describes a balance with a movable fulcrum which was adjusted to restore equilibrium on plunging both the alloy and a silver counterpoise into water. A knowledge of this earlier device was probably the basis of Galileo's neater method. The *Treatise on Ores and Assaying* by Lazarus Ercker (published in German in Prague in 1574) devoted much space to density assays, and there are many descriptions thereafter. Most writers failed to distinguish between the volume fraction of alloy (which varies directly with density) and the weight fraction, which is the figure usually wanted and which varies directly with the specific volume. The two may differ considerably. In gold-silver alloys, for example, the alloy of mean density has 64.8 per cent of gold by weight, while the alloy containing 50 per cent by weight corresponds only to 35.2 per cent by volume.

The density assay was too inexact to be used for more than rough measurements of precious metal alloys. Its most important use was in the control of the tin-lead alloy, pewter, where the economic consequence of error was less. From at least as early as the middle of the fourteenth century the Guild of Pewterers in London controlled the quality of the wares made by its members by casting a ball in a standard mold and comparing its weight with that of a ball cast from a standard alloy in the same mold. Lead was the common adulterant, and if more than the legal amount had been added to the lighter tin, the extra weight was easily detected.

The principle of Galileo's balance is as follows: If W is the weight of an object in air and w is its weight in water, than $w/W = cg/ca$, where cg and ca are the distances from the fulcrum, c, to the position of the counterpoise when balancing the sample in water and in air respectively (see drawing on page 117).

The specific volume of the sample (the reciprocal of the specific gravity) is equal to the loss of weight in water divided by the weight in air, or

$$V = \frac{W - w}{W} = \frac{ca - cg}{ca}$$

A sample of pure gold in water will be balanced by the counterpoise at point e, and pure silver similarly at point f.

The specific volumes of gold and silver are therefore $\dfrac{ca - ce}{ca}$ and $\dfrac{ca - cf}{ca}$ respectively.

The volume of an alloy of gold and silver is practically equal to the sum of the volumes of the component metals, and the specific volume is therefore

$$V = A\left(\frac{ca - ce}{ca}\right) + (1 - A)\left(\frac{ca - cf}{ca}\right)$$

where A and $(1 - A)$ are, respectively, the fractions by weight of gold and silver in the alloy. By equating these two expressions for V and simplifying we see that

$$A = \frac{cf - cg}{cf - ce} = \frac{fg}{fe}$$

i.e., the weight fraction of gold in the alloy is equal to the displacement of the counterpoise point for the alloy from that for pure silver divided by the equivalent distance for pure gold. Galileo gives the ratio of the weights of the two components in the alloy, i.e., $\dfrac{A}{1-A}$, which is equal to $\dfrac{fg}{ge}$.

It is unnecessary to know the actual weight of the sample, since the proportions are all that is necessary, or even to know any dimensions of the balance except *ef* and *eg*. The actual distance corresponding to the entire range of composition between pure gold and pure silver depends, of course, on the length of the balance arm. One equal to a Florentine *braccio* (58.4 centimeters) would require a counterpoise movement of 3 centimeters to rebalance gold in water and 5.6 centimeters to rebalance silver. With the whole distance between the gold and silver points thus 2.6 centimeters, the composition of the alloy could not be very exactly determined, even using the wire-marked subdivisions ingeniously proposed by Galileo. The methods of chemical analysis in use in the sixteenth century would have been much more precise, but they could not have been used without cutting out and destroying a small sample from the object under test.

C. S. S.

BIBLIOGRAPHY

Available English Translations of Galileo's Works

THE ASSAYER. In Stillman Drake, *Discoveries and Opinions of Galileo*. Doubleday Anchor Books, Garden City, New York, 1957.

DIALOGUE CONCERNING THE TWO CHIEF WORLD SYSTEMS. Translated by Stillman Drake. University of California Press, Berkeley, Calif., 1953.

DIALOGUE ON THE GREAT WORLD SYSTEMS. In the Salusbury translation, edited by Giorgio de Santillana. University of Chicago Press, Chicago, Ill., 1953.

LETTER TO THE GRAND DUCHESS CRISTINA. In Stillman Drake, *Discoveries and Opinions of Galileo*. Doubleday Anchor Books, Garden City, N. Y., 1957.

LETTERS ON SUNSPOTS. In Stillman Drake, *Discoveries and Opinions of Galileo*. Doubleday Anchor Books, Garden City, N. Y., 1957.

THE STARRY MESSENGER. In Stillman Drake, *Discoveries and Opinions of Galileo*. Doubleday Anchor Books, Garden City, N. Y., 1957.

DIALOGUES CONCERNING TWO NEW SCIENCES. Translated by Henry Crew and Alfonso de Salvio. Dover Publications, New York, N. Y., 1953.

English Books about Galileo Now in Print

Drake, Stillman, DISCOVERIES AND OPINIONS OF GALILEO. Doubleday Anchor Books, Garden City, N. Y., 1957.

Santillana, Giorgio de, THE CRIME OF GALILEO. University of Chicago Press, Chicago, Ill., 1955.

Books of General Interest

Andrade, E. N. da Costa, SIR ISAAC NEWTON. Macmillan, New York, 1954. Doubleday Anchor Books, Garden City, N. Y., 1958.

Armitage, Angus, THE WORLD OF COPERNICUS. New American Library, New York, N. Y., 1956.

Butterfield, H., THE ORIGIN OF MODERN SCIENCE, 1300–1800. Macmillan, New York, N. Y., 1956.

Cohen, I. Bernard, THE BIRTH OF A NEW PHYSICS. Doubleday Anchor Books, Garden City, N. Y., 1960.

Cooper, Lane, ARISTOTLE, GALILEO AND THE TOWER OF PISA. Cornell University Press, Ithaca, N. Y., 1935.

*Crombie, A. C., AUGUSTINE TO GALILEO—A.D. 400-1650. Harvard University Press, Cambridge, Mass., 1953.

Dampier, W. C., SHORTER HISTORY OF SCIENCE. Meridian Books, New York, N. Y., 1944.

Dreyer, Johan L. E., HISTORY OF ASTRONOMY FROM THALES TO KEPLER. Dover Publications, New York, N. Y., 1953.

Koestler, Arthur, THE SLEEPWALKERS. Macmillan, New York, N. Y., 1959.

——— THE WATERSHED. Doubleday Anchor Books, Garden City, N. Y., 1960.

Kuhn, Thomas S., THE COPERNICAN REVOLUTION. Harvard University Press, Cambridge, Mass., 1957.

Taylor, F. S., ILLUSTRATED HISTORY OF SCIENCE. Praeger, New York, N. Y., 1955.

---

*1959 edition reprinted by Dover Publications, Inc. (0-486-28850-1).

Index